Basic Processes in Physical Geography

Peter M. Eyre and
Graham J. Gower
Watford Grammar School

University Tutorial Press

All rights reserved. No portion of the book
may be reproduced by any process without
written permission from the publishers.

© Peter M. Eyre and
Graham J. Gower, 1983

ISBN 0 7231 0848 X

Published 1983

Published by University Tutorial Press Limited
842 Yeovil Road, Slough SL1 4JQ

Photoset and printed in Great Britain
by Redwood Burn Limited, Trowbridge, Wiltshire
and bound by Pegasus Bookbinding Limited, Melksham

Contents

Introduction

Chapter 1 Processes and Forms in the Crust 1
- The Distribution of Earthquakes and Volcanoes 1
- The Structure of the Earth 3
- Crustal Plates 5
- The Folding and Faulting of Rocks 8
- Rock Types 12
- Extrusive Igneous Activity: Volcanoes 13
- Intrusive Igneous Activity 19
- The History of the Earth 20
- Workshop 23

Chapter 2 Rocks and Relief 24
- Weathering 24
- Slopes 26
- Cuestas and Vales 30
- Rock Types and Relief 35
- Workshop 41
- A Note on Hypothesis Testing and the Scientific Method of Enquiry 42

Chapter 3 Atmospheric Processes 43
- The Weather 43
- Measuring and Recording the Weather 44
- Energy in the Atmosphere and Factors Affecting Temperature 47
- Water in the Atmosphere 52
- Atmospheric Pressure and Winds 59
- Workshop 64

Chapter 4 World Climates, Vegetation and Soils 66
- World Climate 66
- Natural Vegetation 78
- Soils 80
- The Natural Regions of the World 85
- Man's Influence on Climate 94
- Workshop 97

Chapter 5 River Processes and Associated Landforms 98
- River Processes 98
- The Development of a River System 101
- The Drainage Basin 101
- The Work of Individual Streams 109
- Landforms Associated with River Processes 115
- Changes in the River System with Time 124
- Map Extract 1 126
- Map Extract 2 128
- Man's Influence on River Systems 132
- Workshop 133

Chapter 6 Glacial Processes and Associated Landforms 134
- The Formation of Ice 137
- The Effects of Ice Movement 139
- Periglaciation 157
- The Complexity of Glaciation 158
- Glaciation and Man 160
- Workshop 161

Chapter 7 Coastal Processes and Associated Landforms 162
- Wave Action 162
- Marine Erosion 166
- Erosional Coastlines 169
- Marine Deposition 171
- Depositional Coastlines 178
- The Effect of Sea Level Changes on Coastal Landforms 182
- Classification of Coasts 185
- Coastal Processes and Man 186
- Workshop 187

Chapter 8 Desert Processes and Associated Landforms 188
- Desert Weathering 190
- Running Water in the Desert 191
- Wind in the Desert 195
- Deserts and Man 202
- Workshop 205

Additional Map Exercises for the Goring Extract 207

Additional Map Exercises for the Aberystwyth Extract 207

Acknowledgements 207

Index 209

Introduction

Over the past 15 years or so new ideas and techniques in Geography have been filtering downwards to the secondary school level. Many textbooks on the so-called 'New Geography' have appeared on human aspects of the subject, but it seemed to us that there was a need for a textbook in Physical Geography which introduced new techniques and covered processes in a more scientific way.

The book is aimed principally at the 'O' Level courses and as such includes material which traditional syllabuses demand. However, it is hoped that the inclusion of processes will make the book useful for the first-year Sixth as well as perhaps giving a little impetus to syllabus change.

Descriptive text has deliberately been kept to a minimum. The inclusion of questions and exercises within the text is aimed at encouraging pupil involvement. Different methods of enquiry like hypothesis testing and correlations are introduced. There are many data response exercises based upon data given in the book, but pupils should have ready access to a good secondary school atlas. Some of the questions are for the lower ability pupil, but others are rather more difficult and may well need the help and guidance of the teacher in answering them. It is hoped that the layout of the book will enable a wide range of different teaching methods to be used.

P.M.E.
G.J.G.

Processes and Forms in the Crust

In the autumn of 1980, two devastating earthquakes hit the Mediterranean region. One destroyed El Asnam in Algeria, a city the size of Oxford. The other laid waste a large area around Potenza in southern Italy. Earthquakes are well known in the Mediterranean region: Skopje, Yugoslavia, was destroyed in 1963 killing over 1000 people, and in eastern Turkey nearly 2500 were killed four years later.

Many earthquakes are recorded every year around the world, but only those that cause damage and destruction to towns and villages actually hit the headlines. Volcanoes receive a good deal of publicity, especially when they explode with the force of a ten megatonne bomb as Mount St. Helens did in May 1980, or when they create a completely new island as at Surtsey, off Iceland, 1963–67.

As with many features in physical geography, we need to study the **distribution pattern** of earthquakes and volcanoes in order to understand how they arise. In understanding their cause, scientists are in a better position to predict when they will occur. Then the terrible effects can be lessened by evacuating people and livestock away from the danger zones.

Volcano	Location
Mt. Wrangell	61° 40′N 143° 0′W
Katmai	58° 17′N 154° 56′W
Shasta	41° 20′N 122° 20′W
Baker	48° 50′N 121° 49′W
Mt. St. Helens	46° 12′N 122° 11′W
Paricutin	19° 28′N 102° 15′W
Popocatepetl	19° 10′N 98° 40′W
Cotopaxi	0° 30′S 78° 30′W
Volcán Corcovado	22° 57′S 43° 13′W
Mauna Loa	21° 08′N 157° 13′W
Asama Yama	36° 24′N 138° 31′E
Fuji-San	35° 22′N 138° 44′E
Kokuritsu-Koen	31° 50′N 130° 55′E
Krakatoa	6° 07′S 105° 24′E
Ruapehu	39° 18′S 175° 35′E
Mt. Erebus	77° 32′S 167° 09′E
Etna	37° 46′N 15° 00′E
Vesuvius	40° 49′N 14° 26′E
Stromboli	38° 48′N 15° 13′E
Surtsey	63° 16′N 20° 32′W
Tristan da Cunha	37° 15′S 12° 30′W
Elgon	1° 08′N 34° 33′E
Kilimanjaro	3° 04′S 37° 22′E

Table **1.1** Major volcanoes

The Distribution of Earthquakes and Volcanoes

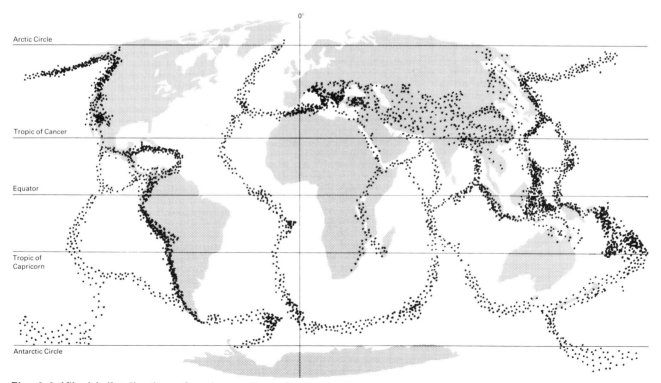

Fig. **1.1** World distribution of major earthquake zones

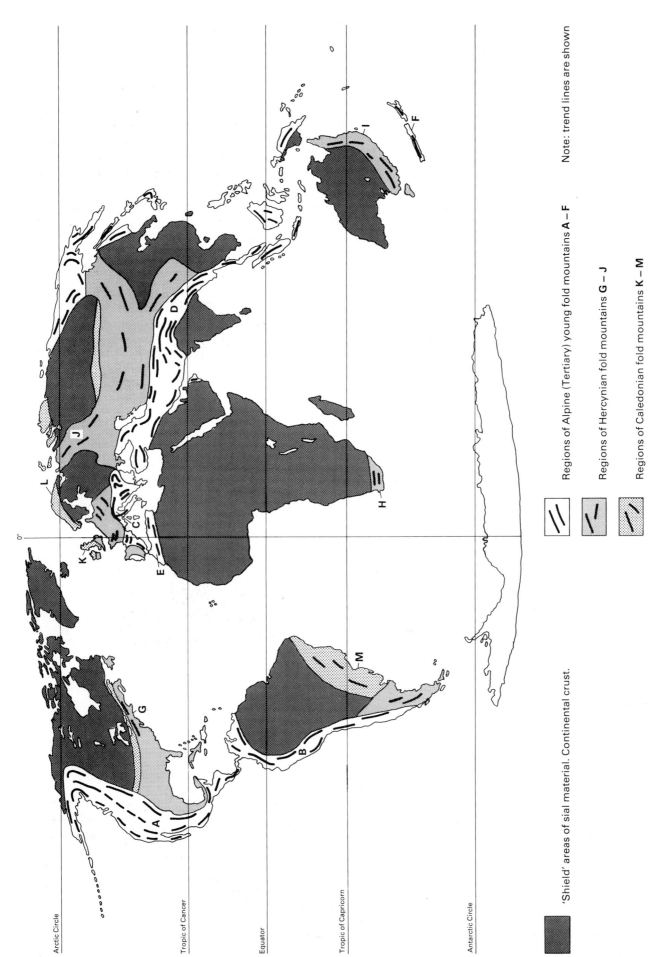

Fig. 1.2 World distribution of fold mountains and continental crust areas

Processes and Forms in the Crust

> **1 a)** Place a piece of tracing paper over the world map, Fig. 1.1. With the help of your atlas, plot the volcanoes listed in Table 1.1 as accurately as you can.
>
> **b)** Describe the relationship between the distribution of the volcanoes on your map and the distribution of earthquakes shown in Fig. 1.1.
>
> **2** Study a map of world relief in your atlas. From Fig. 1.2 identify the ranges of **young fold mountains**. How is the distribution of these related to the distribution of volcanoes and earthquakes?

Earth is constructed and to the processes that go on inside the structure.

The Structure of the Earth

Fig. 1.3 is a model showing how geophysicists think the Earth is constructed. The information for this is obtained from a number of different sources, but particularly from the patterns made by earthquake shock waves as they pass through different types and densities of material.

The **crust** of the Earth is so thin that we need to enlarge the scale considerably to see how it is made, as in Fig. 1.4. In fact, if the Earth were reduced to the size of an apple, the crust would be thinner than the skin, and the Himalayas would hardly show as wrinkles.

We have now seen how the important physical phenomena of the world occur together in belts. Their distribution patterns are said to be **correlated**. This important correlation is due to the way in which the

> **3** Study Fig. 1.4 carefully. What zones can the crust be divided into? What is the approximate thickness of each of these zones?

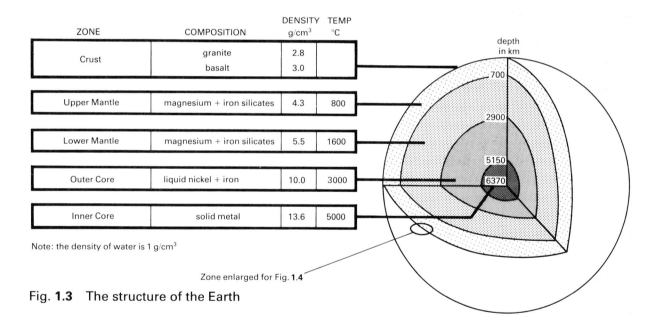

Fig. **1.3** The structure of the Earth

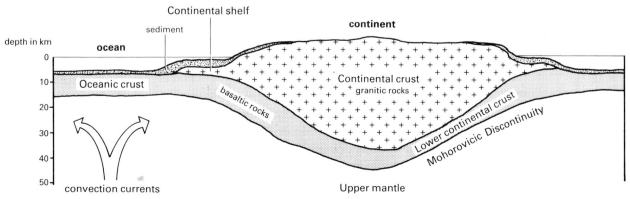

Fig. **1.4** The structure of the crust

Note: the vertical scale is exaggerated considerably

The **continental crust** is made up of **granitic rocks** which consist largely of complicated aluminium silicates (silica is an oxide of silicon). This part of the crust is often termed the **sial** (*si*licon + *al*uminium) layer because of its composition (see Table 1.2 on page 12).

On top of the continental crust is a very thin veneer of **sediments**. These are produced by the breakdown of the granitic rocks under the effects of the weather (see page 24).

> **4** On what part of the continental crust are the sediments the thickest?

The **oceanic crust** is made up of **basaltic rocks** which consist of the silicates of iron and magnesium. It is often called the **sima** (*si*licon + *i*ron + *ma*gnesium) layer. It extends below the continental crust as the lower continental crust, although the chemical and physical composition may be changed by the weight of the material above it.

The crust represents the lightest material of the Earth. It settled out on the surface of the planet after its formation from a mass of gas some 4600 million years ago. As it cooled, the crust solidified. It can be described as *floating* on the **mantle** beneath.

The **upper mantle** is very hot and behaves more like a liquid than a solid. It is thought to have **convection currents** rising through it: these have a very important role in the formation of the features of the crust. Materials will expand as they are heated; as they expand, they become less dense and lighter. The mantle material is hot near the core of the Earth. It expands, becomes less dense, and rises towards the surface. As it does so, it cools down because it reaches a cooler zone. This cooling causes the material to contract, become denser and heavier, and it spreads horizontally and descends to be heated again. This results in a roughly circular movement of mantle material, known as convection.

A model of this can be made in the chemistry laboratory. Take a flask of water. Drop a small piece of potassium permanganate onto the bottom of the flask. Place the flask over a bunsen flame and observe the patterns made by the purple streaks as the water heats up. These patterns are, in simple form, similar to those that may exist beneath the crust (Fig. 1.5).

> **5** Study the information given in Figs. 1.3 and 1.4.
> **a)** Why has the crustal material risen to the surface?
> **b)** Why do you think that the crustal material split into the continental and oceanic layers?
> **c)** **Mohorovicic** was a Yugoslavian geologist who discovered an important boundary, or **discontinuity.** Where is this?
> **d)** What do you notice about the thickness of the continental crust above and below sea level?

Fig. **1.5 (a)**   Convection currents in a flask

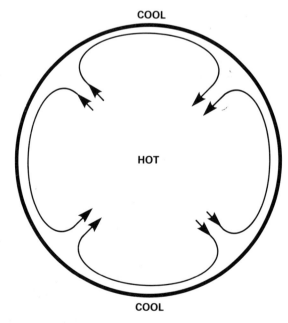

Fig. **1.5 (b)**   Convection currents in the Earth. (This is purely diagrammatic and should not be taken to mean that there are four currents under the crust.)

Processes and Forms in the Crust

The answer to the last question has a lot to do with balance. Icebergs have 90% of their mass below sea level so that they float in a stable manner without toppling over: the same principle applies to the continents, and is known as **isostasy**. The mass of the mountain chains extends downwards in order to maintain balance: this mass is known as the **roots** of the mountains.

Crustal Plates

We have seen that the crust is made of a thin layer of granite (the continental crust) and basalt (oceanic crust) which floats on the mantle beneath. However, the crust is not a continuous, unbroken layer: it is split into a number of segments known as **plates**. Fig. 1.6 shows the distribution of these plates and the boundaries along which they meet. The cross-section in Fig. 1.7 shows their distribution around the Equator.

> **6** Describe any correlations you see between the pattern made by the plate boundaries and the distribution of earthquakes, volcanoes and young fold mountains.

These plates are not stationary but are moving slowly over the mantle in the directions shown by the arrows in Figs. 1.6 and 1.7. The speed of movement varies from place to place, but it is estimated that the American Plate and the Eurasian Plate are drifting apart at the rate of 1 cm a year.

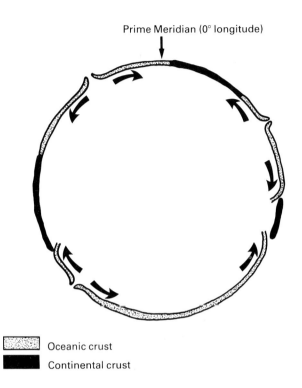

Fig. **1.7** Cross-section through the Equator showing distribution of plates

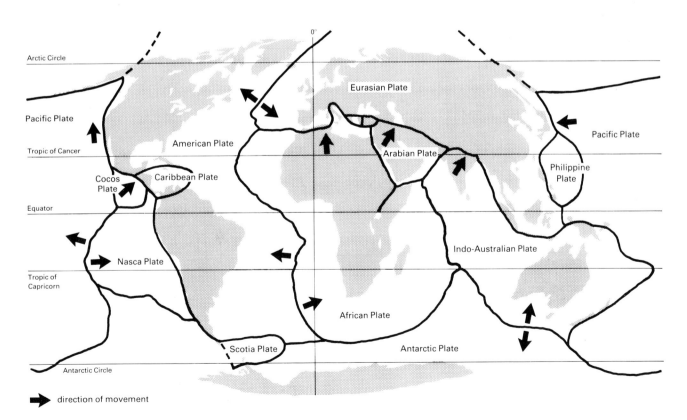

Fig. **1.6** The distribution of crustal plates

7 Study together Figs. 1.6 and 1.7. Moving eastwards from the Prime Meridian (0° longitude), identify the continental crustal plate sections and the oceanic crustal plate sections on a copy of Fig. 1.7.

Between the ridges on either side of the gap, the mantle material cools as it spreads. This cooling causes changes in the chemical composition of the rock, which solidifies to become new oceanic crust. The whole process is called **sea floor spreading**. More explosive mantle activity causes the formation of volcanoes which may protrude above the surface of the ocean as islands.

Plates which Move Apart (Divergent Plates)

Fig. 1.8 shows in diagrammatic form what is thought to be happening on the bed of the mid-Atlantic. Here, the hot mantle material is rising upwards and spreading outwards, causing the crustal plates to move apart and the edges to buckle upwards into a series of **ridges**.

8 Identify as many volcanic islands as you can in the mid-Atlantic, using your atlas.

9 Find other mid-ocean regions that also have ridges as a result of neighbouring plates moving apart.

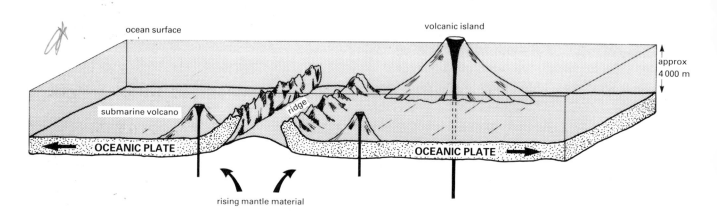

Fig. **1.8** A diagrammatic cross-section through the mid-Atlantic ridge

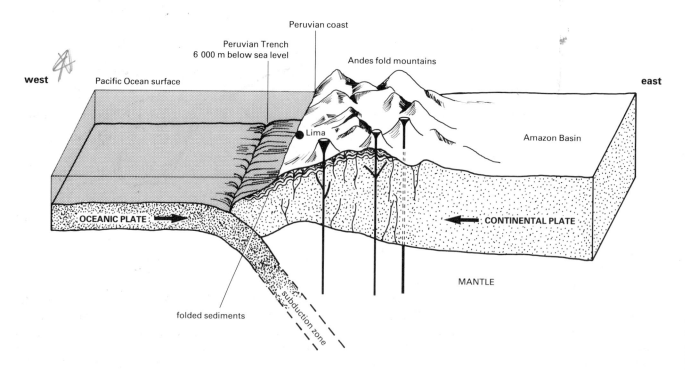

Fig. **1.9** A diagrammatic cross-section through the west coast of South America

Processes and Forms in the Crust

Plates which Move Together (Convergent Plates)

10 Fig. 1.9 shows what is thought to be happening along the coast of Peru on the western side of South America. Name the two plates involved.

The oceanic plate is forced down into the mantle, where it breaks up and melts in what is known as the **subduction zone**. Masses of molten oceanic material well upwards through the mantle.

11 Why should the oceanic material melt in the subduction zone? Explain why it rises upwards (Fig. 1.3 gives a clue).

12 When this molten material finds its way through cracks and weaknesses in the continental crust above, what feature may be formed at the surface?

The movement of the plate in the subduction zone is neither smooth nor regular. It can in fact be sharp and violent, sending shock waves through the Earth: these are **earthquakes**. Earthquakes also occur as the continental crust, being lighter than the oceanic crust, rides above the plate boundary. The granitic rocks and the sediments get crumpled and folded. This contortion and uplift creates chains of fold mountains. A similar situation to that of the west coast of South America occurs along the east coast of Asia, where the Pacific Plate is dipping under the Eurasian Plate.

13 Refer to Fig. 1.30 on page 22. How young are the young (Alpine) fold mountains?

These mountains are also referred to as **Tertiary** fold mountains because they were built up in the Tertiary era of the Earth's history. Fig. 1.30 shows that they are also the youngest. Mountain building is called **orogenesis**, and has occurred several times during the Earth's evolution.

Looking at Fig. 1.4 you will notice that the continental plate does not end at the coast but extends some way beneath the ocean as the **continental shelf**. These shelf regions have very important implications for mankind. As the ocean is relatively shallow, sunlight can penetrate, encouraging marine life: continental shelves are the world's most valuable fishing grounds. They are also potential sources of minerals: North Sea oil is a good example of a shelf resource which is already being exploited.

14 Find a physical map of the world in your atlas. Where are the widest continental shelves to be found?

Where plates are moving together, the continental shelf has an outer margin that dips down steeply to a **marine trench**.

15 How deep is the Peruvian trench? (Fig. 1.9). Find in your atlas where the deepest marine trenches are to be found.

Plates which Move Sideways (Transform Plates)

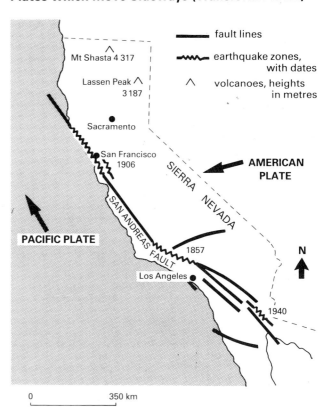

Fig. **1.10** The San Andreas fault zone

Along the western coast of North America, the Pacific Plate is moving northwards in relation to the American Plate, along the San Andreas Fault. This movement is being studied closely in California because it could cause an earthquake like the one that led to the devastation of San Francisco in 1906. The complexity of faults (lines of weakness in the crust) in this region tends to lock the plates together, preventing them from moving smoothly. Stress and tension build up in the crust which can suddenly give way, causing severe earthquakes. In 1864, 1898 and 1900 there were severe quakes, but the greatest so far occurred at 5.13 a.m. on 18 April 1906. The earthquake itself did less damage than the fire which followed it. Up to 700 people are thought to have died and 250 000 were made homeless. Ten square kilometres of the city centre were completely destroyed. However, the local distillery survived, which prompted a newspaper reporter from the East Coast to write:

> 'If, as they say, God struck the town
> For being over frisky,
> Why did He burn the churches down
> But spare Hotaling's whisky?'

Seismologists (those who study earthquakes) and geologists warn that another earthquake can be expected with reasonable certainty before the end of the century, although it is very difficult to accurately predict such things. With the number of modern tower blocks that now dominate San Francisco, the next quake could have far more disastrous effects than the one in 1906.

Tokyo is another city that has been badly damaged by an earthquake in recent times, in 1926. It lies on the margin of the Eurasian and Pacific Plates. In this case the mechanics of the cause of the quake are slightly different. Read again page 7 to see why.

The situation on the west coast of North America is made more complicated by the fact that not only is the Pacific Plate moving laterally northwards, but the American Plate is moving westwards against it. This is causing the rocks of the continent to buckle and fold, and volcanoes are forming in a similar way to those along the west coast of South America.

16 It has been said that one day Los Angeles might lie opposite Sacramento. If the average rate of movement of the Pacific Plate is 5 cm a year, calculate approximately when this might happen (Fig. 1.10).

The Folding and Faulting of Rocks

We have seen that the rocks of the continental crust become contorted and crumpled at the plate boundaries, the shock waves that result from these movements causing earthquakes. Distortions of rocks can be divided into **folding** and **faulting.** Figs. 1.11 and 1.12 illustrate the simplest kinds of folding and faulting, but it must be realised that such features occur in complex arrangements and in a wide range of scales from the immense folds of the Alps to the merest buckle and crack in a coal seam only centimetres across.

17 From the information in Fig. 1.11, describe in your own words the following terms: **anticline, syncline, crest, trough, dip, strike** and **limb.**

Fig. **1.11 (a)** Folding in sedimentary rocks, Lulworth Cove, Dorset

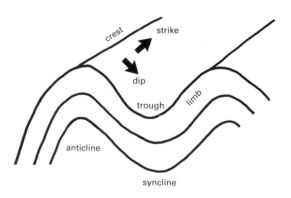

Fig. **1.11 (b)** Simple folding in rocks

Processes and Forms in the Crust

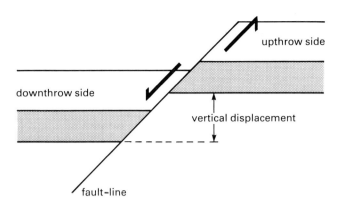

Fig. **1.12** Cross-section through a normal fault

Simple folding can cause gentle relief with hills on the crests and valleys in the troughs. Such an arrangement can be seen in the Jura Mountains on the border between Switzerland and France. These were folded about 50 million years ago during the formation and uplift of the great Alpine mountain chain. Folding in the Alps is extremely complicated. Fig. 1.13 gives only a very simplified version of it. A variety of different types of fold are seen in great fold mountain chains like the Alps, the Himalayas and the Andes. Fig. 1.14 illustrates the types of folding that may be found. Notice that a **nappe** also involves faulting as a mass of rock is thrust horizontally.

In Fig. 1.13 you can see that the surface relief bears little or no resemblance to the folding beneath. This is because the surface has been severely eroded (worn away) by the action of mountain glaciers (see Chapter 6).

The Faulting of Rocks

The San Andreas Fault is a very long **lateral fault** at the junction of two plates, but faulting on a smaller scale occurs frequently within plates themselves. Apart from sideways, faulting displaces rocks upwards or downwards along a crack, or **fault-line**, in the crust.

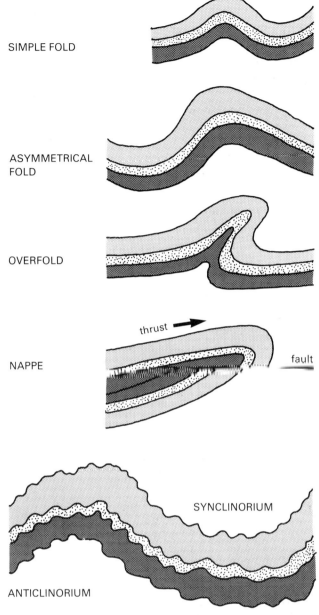

Fig. **1.14** Types of folding

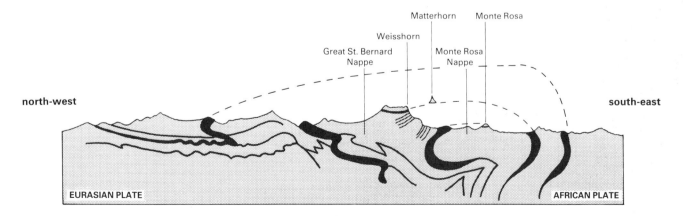

Fig. **1.13** Complex folding in the western Alps. Broken lines indicate rock masses that have been eroded away. The Matterhorn lies off the line of section to the south.

Fig. 1.15 is a photograph of a fault in a quarry in the Forest of Dean, Gloucestershire.

Fig. **1.15** A reverse fault, Forest of Dean

18 Draw a sketch of this photograph and label it in a similar way to the fault shown in Fig. 1.12. What is the difference between the two faults illustrated?

Fig. 1.12 is an example of a **normal fault**, where the surface has been lengthened. Fig. 1.15 is an example of a **reverse fault**, where the surface has been shortened.

19 Measure the amount of vertical displacement in Fig. 1.15.

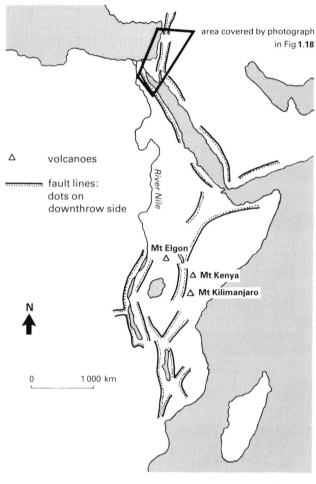

Fig. **1.17** The East African rift system

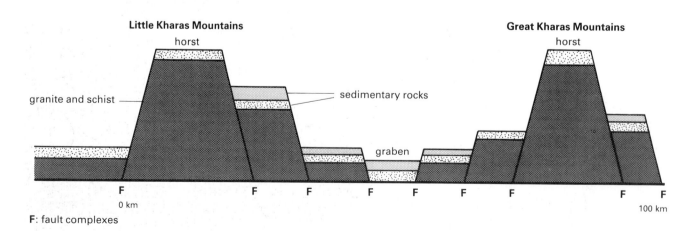

Fig. **1.16** Simplified cross-section through the Kharas mountains, Namibia

Faults often occur in groups with blocks being thrust upwards to form **horsts** or downwards to form **graben**, as shown in Fig. 1.16. Elongated graben form **rift valleys**, such as the Rhine Rift Valley between Mainz and Basle.

Fig. 1.17 shows a rift system on a scale similar to that of the San Andreas Fault. Fig. 1.18 shows a satellite photograph of part of the system.

> 20 Locate the Rhine Rift Valley in your atlas.
> a) What is its length and average width?
> b) The hills to the east and west at the southern end are examples of horst blocks: name them. How high are they above the valley floor?
> c) Draw a profile through the valley at Colmar. The boundaries between the shades on your atlas map are the contour lines. Label the Rhine and the hills to either side.

> 21 Make a tracing of Fig. 1.18 and mark on it the plate boundary (see Fig. 1.6), the names of plates 1 and 2 and sea areas A, B, C, and D.

The section in East Africa itself is thought to be due to the plate splitting under internal tensions. Note that volcanoes are found in this region. The Rhine Rift Valley at one time also had volcanic activity, but the volcanoes are now extinct, e.g. the Krehberg, 35 km north-east of Mannheim.

Fig. **1.18** Satellite photograph of the Sinai Peninsula, part of the East African rift system

Rock Types

Igneous Rocks

Rocks are aggregates, or mixtures, of **minerals**. Minerals are composed of the atoms of chemical **elements**. There are about 92 elements, but fortunately the most common rocks can be described in terms of a dozen or so.

Name	Symbol	Percentage
Oxygen	O	46.71
Silicon	Si	27.69
Aluminium	Al	8.07
Iron	Fe	5.05
Calcium	Ca	3.65
Sodium	Na	2.75
Potassium	K	2.58
Magnesium	Mg	2.08
Titanium	Ti	0.62
Hydrogen	H	0.14

Table **1.2** The main chemical elements which make up the crustal rocks

Oxygen combines readily with the other elements to form oxides, the most common being **silica** (silicon oxide), SiO_2. This mineral is very common in rocks and is known as **quartz**. **Sand** is made up of quartz grains. Silica in turn combines with other elements to form silicates, the most common being the **felspar** group made up of potassium, aluminium and calcium silicates. These are important constituents of granite.

During the time when the Earth was a molten ball, the lighter elements floated to the surface. As the surface cooled, so the elements combined and crystallised into minerals to make up **igneous rocks**: granite of the continental crust, basalt of the oceanic crust. Volcanic outpourings throughout geological time have also contributed to the Earth's stock of igneous rocks. Hot gases and liquids crystallising within the crust have given us important mineral resources of the precious and semi-precious type as well as non-ferrous metals, e.g. gold, silver, tin, lead, zinc and copper.

Granite consists of three major mineral groups: felspar, mica (both of which are silicates of aluminium, potassium and calcium) and quartz. If you have access to a piece of granite, try to identify these minerals. Quartz is glassy, felspar creamy and in some cases pink, and mica is black or silver. The grains are of measurable size, often up to several millimetres across.

Basalt is a much finer grained rock and is much darker in colour due to the presence of minerals containing iron and magnesium, such as the **pyroxene** group and biotite (a form of mica).

Granite and basalt are thus very different in content and appearance. One major difference lies in the amount of quartz. Granite has a lot of quartz and as such is classified as an **acidic** igneous rock. Basalt, on the other hand, has little or none, and is classified as a **basic** igneous rock. This difference has important effects upon the landforms they produce. Between granite and basalt is a wide range of **intermediate** rocks whose chemical composition is intermediate between acidic and basic.

The difference in grain size between granite and basalt is the result of different rates of cooling of the molten rock, or **magma**. Magma deep in the crust cools slowly, producing large grains. Such rocks are sometimes referred to as **intrusive** or **plutonic**. On the surface, the material will cool quickly, resulting in small grains which can be so minute that they are invisible to the naked eye. Such rocks are called **extrusive** or **volcanic**.

Igneous rocks form a complex group, but matters can be made easier to understand if you remember the simple two-fold classification: a) acidic or basic b) extrusive or intrusive.

---

22 Find a geological map of Britain in your atlas.
a) On an outline map, shade on the distribution of igneous rocks. Pay attention to the way in which the rocks have been classified.
b) Describe the distribution you have mapped. What is the relationship between the rocks and the height of the land?

Sedimentary Rocks

Igneous rocks exposed on the surface will be broken down both physically and chemically by the weather elements (weathering) and worn away by streams, the sea and glaciers (erosion). The weathered and eroded material will be deposited on the floors of lakes, on river flood plains and, most of all, on the bed of the sea. These **sediments** will build up in layers, or **strata**, and become compressed and cemented to form **sedimentary rocks**. The quarry face shown in Fig. 1.15 clearly shows sedimentary rock strata. The strata are also called **beds**; the boundary between each bed is called a **bedding plane**.

The composition of the sedimentary rock depends very much upon the nature of the original igneous rock. Granite decomposes to form **sand** (the quartz grains) and **clay** (from the decomposition of felspars). Basalt produces clay as well as many soluble (dissolvable) products. Sand grains cemented together by silica or calcite form **sandstones**, which tend to be hard and resistant to weathering and erosion. Clay, on the other hand, and **shale** (a coarser kind of clay) are soft.

The sea is home to millions of creatures with shells made of calcium carbonate. These accumulate to considerable depths as the creatures die and sink to the

Processes and Forms in the Crust

bottom. Such accumulations give rise to a group of sedimentary rocks known as **limestones**. **Chalk** is a very pure form of limestone consisting of microscopic shells. In many chalk deposits, masses of **flint** are found, either in layers or in rounded masses called nodules. This flint is thought to be the result of the chemical deposition of silica. Some limestones are also the result of chemical, rather than organic, deposition. They are often hard and massive, that is, made up of very thick beds.

> **23 a)** Shade on your outline map from exercise 22 the areas of chalk, limestone, sandstone, clay and shale.
> **b)** Describe the distribution shown. Describe the relationships between the height of the land and the rock types. How might you account for these relationships?

Sedimentary rocks are themselves weathered and eroded and the resulting material is formed into new sedimentary rocks.

Metamorphic Rocks

All rocks, whatever their origin, may at a later date be altered by intense heat and pressure from crustal movements or by igneous intrusions. The changed rocks are called **metamorphic rocks**. Clays and shales can be compressed into **slates**, granite is metamorphosed into **gneiss**. Metamorphic rocks tend to be very hard and resistant.

> **24** Shade on your geological map of Britain (exercise 22) the distribution of metamorphic rocks. What reasons can you put forward for this particular distribution pattern?
>
> **25** What particular area is famous for its slate? To what uses is slate put?

Extrusive Igneous Activity : Volcanoes

Volcanoes provide some of the most dramatic features on the Earth's surface. They can be very destructive as in the case of Vesuvius in A.D. 79 when ash buried the city of Pompeii. However, in the early stages of the Earth's history, steam and gases from volcanoes

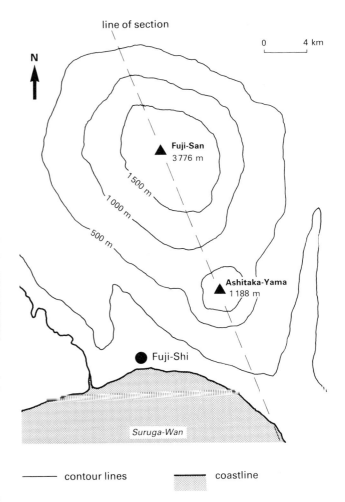

Fig. **1.19 (a)** Fuji-San, Japan

provided our atmosphere and oceans. The rich soils on the flanks of many volcanoes, including Vesuvius, are used for agriculture.

Figs. 1.19 (a) and (b) are maps of two very different types of volcano, Mauna Loa on Hawaii, and Fuji-San in Japan.

> **26** Draw a profile (section) of Mauna Loa and Fuji-San along the lines indicated on the maps. Allow for the different scales of the two maps.

It is apparent from the profiles that one volcano is steeper than the other.

> **27** For each volcano calculate the average gradient of its slope along the line of section. (By the *average* gradient we mean that all the irregularities are ignored.)

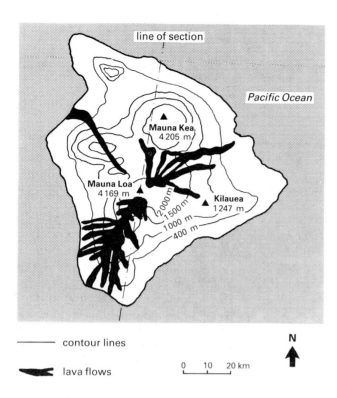

Fig. **1.19 (b)** The island of Hawaii

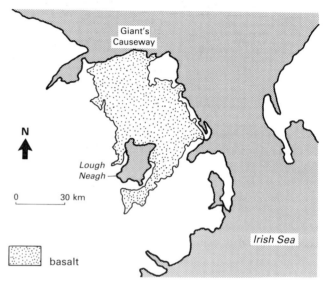

Fig. **1.20** Tertiary plateau of basalt in Antrim

to create **basalt plateaus.** During the Tertiary era, 30 to 50 million years ago, the area between Iceland and N.W. Britain was covered by one such plateau. Much of it foundered beneath the Atlantic during later plate movements, but remnants exist in west Scotland and in Antrim, Northern Ireland (Fig. 1.20).

The difference shown in your answer is explained by the difference between the two types of **lava** that are emitted during an eruption. The lava in Mauna Loa is of a basic variety, containing little or no silica. This allows it to flow very freely over long distances. In the case of Fuji-San, the lava contains a good deal of silica. It is an acidic lava that does not flow very easily and which builds up steep-sided layers.

These two volcanoes illustrate two fundamental types: the **Hawaiian type** (or **shield** volcano) and the **Vesuvian type** (Vesuvius having a similar shape to Fuji-San). There is a third type which is recognised – one that does not have any lava emission at all, known as the **Peléean type** (after Mont Pelée on the island of Martinique, see page 17).

Basic free-flowing lavas have a temperature of over 1000°C. They can spread over hundreds of square kilometres, sometimes from not just one crater but from several dispersed along a fissure (crack) in the crust, as in the Laki fissure in Iceland. Activity is reasonably quiet and gentle with reservoirs of heaving lava inside the craters.

Mauna Loa is spectacular in producing 'fire fountains' of spraying liquid rock. This volcano is actually much larger than it appears to be on the map. Fig. 1.19 (b) shows only the part that emerges above the surface of the Pacific. Altogether it rises 10 000 m from the ocean bed (making it taller than Everest) with a base at least 450 km in diameter.

Some outpourings from fissures have been even more fluid, with lava covering thousands of square kilometres

The cooling of the basalt in Antrim caused the rock to contract into regular polygonal columns, many being hexagonal: these are well exposed on the coast at the famous Giant's Causeway.

Other extensive basaltic plateaus occur in the Columbia–Snake River region of the north-west USA, and in the Deccan of India which covers 640 000 km^2 up to a depth of 2000 m in places.

28 What is the approximate area covered by the basalt in Antrim, shown in Fig. 1.20? A precise method of measuring irregular areas is outlined on page 107.

The Japanese regard Fuji-San as sacred, partly because of its almost perfectly symmetrical shape. However, your profile of the mountain shows a marked irregularity, formed by the small cone of Ashitaka-Yama. This is known as a **parasitic cone**, fed from a subsidiary feed pipe. Fig. 1.21 is a diagrammatic section through a model volcano and represents what the structure of Fuji-San and similar volcanoes are like inside.

Processes and Forms in the Crust

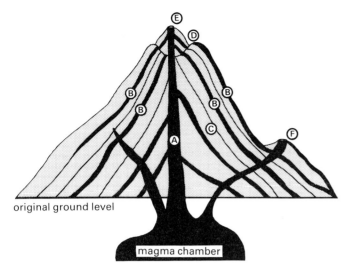

- Ⓐ feed pipe or conduit
- Ⓑ lava flows
- Ⓒ layers of pyroclastic material
- Ⓓ explosion crater
- Ⓔ eruptive cone
- Ⓕ parasitic cone

Fig. **1.21** A diagrammatic cross-section through a composite cone

29 Describe the structure of the model. (Note: **pyroclastic material** is the accumulation of pulverised rock.)

Because of their structure, such cones are called **composite cones.**

The lava is very viscous, requiring great force to eject it from the feed pipe. If the lava solidifies in the pipe, the build-up of pressure from below can be enormous, resulting in an explosion which can blow the top off the mountain. Such a **paroxysmal eruption** occurred at Mount Katmai, Alaska, in 1912. The original summit reached 2300 m but on 6 June two explosions lowered the top by 300 m. The cone was replaced by a crater, 5 km across and 800 m deep, in which a small cone of cinders formed an island. The much enlarged crater is known as a **caldera.** Perhaps the most famous caldera of all is Crater Lake in Oregon which is nearly 10 km across. However, Crater Lake was not formed by a paroxysmal eruption but by **cauldron subsidence.** The large magma chamber beneath the mountain partially emptied during an eruption 6500 years ago. The roof of

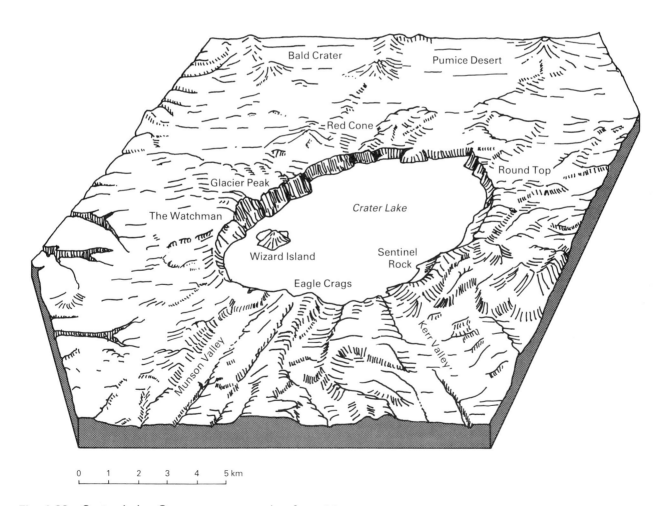

Fig. **1.22** Crater Lake, Oregon, an example of a caldera

the chamber could no longer support the massive weight of the mountain above it, so it collapsed, causing the caldera to form. This caldera subsequently filled with water, to form a lake.

> **30** Fig. 1.22 shows Crater Lake with Wizard Island. What do you think is the probable origin of Wizard Island?

Magma rises through fissures in the crust to produce volcanic activity, firstly because pressure above it is reduced along the line of the fissure, and secondly because it is infused with gases, rather like soda water is infused with carbon dioxide. These gases include steam, carbon dioxide and the compounds of sulphur and hydrogen. The magma becomes lava as the gases are given off, rather like soda water becoming ordinary water as the carbon dioxide is released. However, some

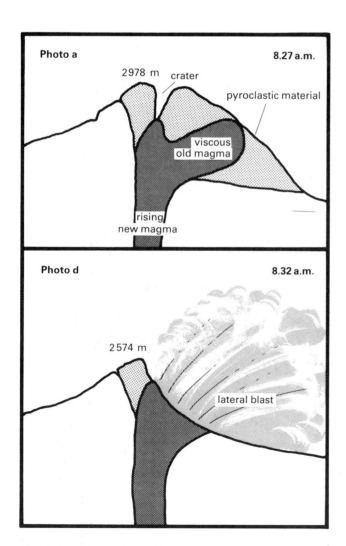

Fig. **1.23** The structure of Mount St. Helens during the eruption of 18 May 1980. (View from N.E.)

Fig. **1.24** The eruption of Mount St. Helens, 18 May 1980. (View from N.E.)

Processes and Forms in the Crust

volcanoes only emit the gases and not the liquid material: they have no lava flows. These are the Peléean type volcanoes.

In 1902, Mont Pelée on the West Indian island of Martinique erupted with a great cloud of gas and incandescent rock fragments (a **nuée ardente**). This rolled down the mountain, causing great loss of life in the port of St. Pierre. According to eyewitness accounts, the nuée ardente caused the sea to boil as it entered.

The most recent and best documented eruption of this type was that of Mount St. Helens which blew up on 18 May 1980. Fig. 1.23 shows the structure of the volcano, and the photographs in Fig. 1.24 show the sequence of events.

31 Describe the changes in the shape of the mountain before and after the eruption. How can you explain these changes?

It has been estimated that over 400 million tonnes of ash and dust were released in a blast equivalent to the force of a ten megatonne bomb. Clouds of superheated gas and ash rolled down the mountainside, killing sixty-one people and an estimated two million birds, and destroying thousands of hectares of valuable forests.

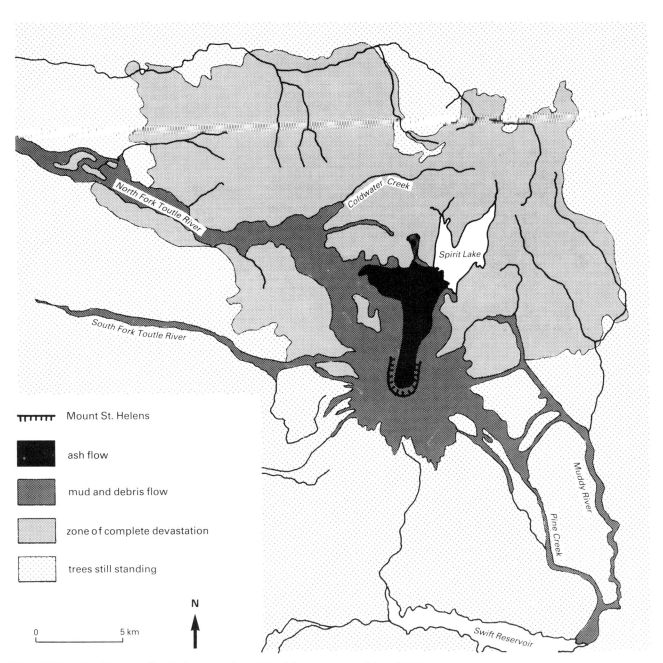

Fig. **1.25** The Mount St. Helens region, Washington, late May 1980

> **32** Study the map shown in Fig. 1.25. What seems to have controlled the distribution of ash, mud and debris flow?

Huge quantities of ash were deposited over a wide area after the eruption. It is estimated that the area covered in ash that was at least 2 mm deep was 260 000 km². The spread of ash into the atmosphere from eruptions like that of Mount St. Helens may have far reaching effects. That which settles can bury crops and homes. That which stays in the atmosphere can cut down the amount of energy received from the Sun, so affecting the climate. Many will remember the summer of 1980 which followed the eruption of Mount St. Helens as being particularly cool, but of course cool summers occur even when volcanoes do not blow up! In Chapter 6 the possible effects of volcanic dust will be discussed in relation to the cause of Ice Ages.

Even as a volcano is being formed it will be attacked by weathering and erosion. The effects these have on the form of the volcano will depend upon the nature of the weathering and erosion agents themselves, the kinds of material that make up the volcano and whether the cone is active, dormant or extinct. A **dormant** volcano is one that has erupted in historical time but not in living memory.

Cones that are made up of **pumice** (solidified lava through which escaping gases have made many pore spaces) and loose ash allow rainwater to pass through them very quickly (that is, they are very **permeable**). This means that surface streams cannot become established and so cannot erode the cone. Some cones on Hawaii are very well preserved, despite being in a climate with an annual rainfall of over 5000 mm. Many cones are high enough for glacial erosion (Chapter 6) to affect their shape, even in the Tropics, e.g. Mt. Chimborazo in Ecuador (6262 m high) is being eroded by glaciers near the peak. With active volcanoes, eruptions can signifiicantly affect the work of streams by filling in valleys with lava or ash and diverting their courses. So a volcano must be regarded as a dynamic landform that can be radically altered in a very short space of time, or slowly over a very long period.

Fig. 1.26 shows the wearing away of a model volcano until all that is left is a stump representing the **plug**: the solidified feedpipe.

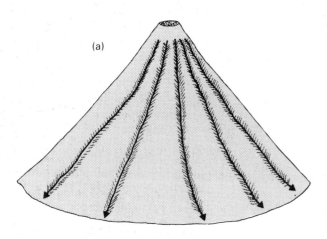

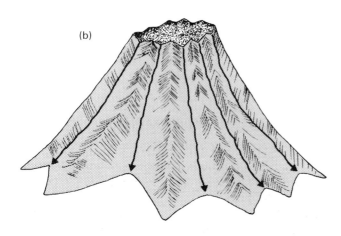

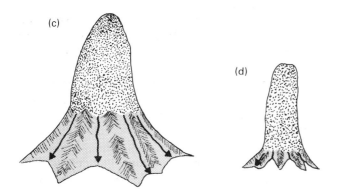

Fig. **1.26** Stages in the erosion of a composite volcano

> **33 a)** Describe the shape of the volcano at each stage shown. Why do you think that only the plug is left? What do you think will eventually happen to the plug itself?
> **b)** Refer to Fig. 1.27. What stages in the model do you think photos (a) and (b) illustrate?

Fig. **1.27 (a)** Volcano de Fuego, Guatemala

Fig. **1.27 (b)** Le Puy, Auvergne, France

Intrusive Igneous Activity

Masses of magma are injected into rock strata and may not reach the surface at all unless exposed by the erosion of the rocks above. Intrusive magma occurs in the following formations:

a) A **dyke** (or dike). A vertical or steeply inclined sheet of igneous rock that has forced its way through a fissure or melted a path for itself across the planes between the rock strata.

b) A **sill**. A sheet of igneous rock that has been injected between the rock strata. Its thickness, which may vary from between a few centimetres to several metres, is insignificant in comparison with its area.

c) A **batholith**. A mass of igneous rock of very great volume extending to considerable depths within the crust. It replaces existing rocks rather than displacing them. The surrounding rocks are metamorphosed by the hot magma before it cools, forming a **metamorphic aureole**. **Stocks** and **bosses**, 50 kilometres or so across, are small versions of batholiths.

d) A **laccolith**. A large dome-shaped mass that has forced its way between strata. The land surface can be domed upwards as a result, forming a large hill.

34 Fig. 1.28 (a) to (e) shows examples of intrusive magma features. Study them carefully, and say which example represents which feature. Give reasons for your choices.

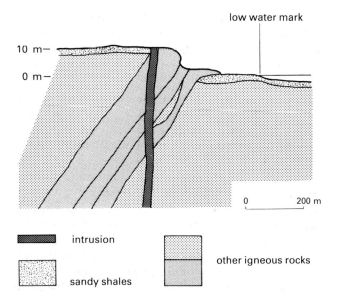

Fig. **1.28 (a)** The shore south of Kilchoan, Ardnamurchan, W. Scotland

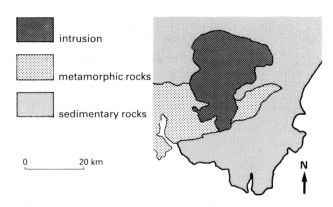

Fig. **1.28 (b)** Dartmoor

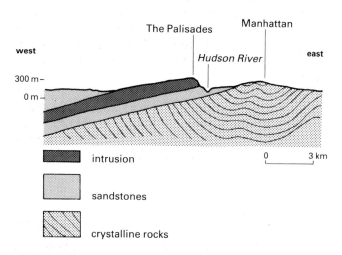

Fig. **1.28 (c)** The Palisades, New Jersey, USA

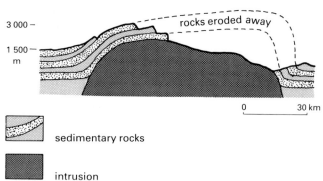

Fig. **1.28 (d)** Mount Hilliers in the Henry Mts, Utah, USA

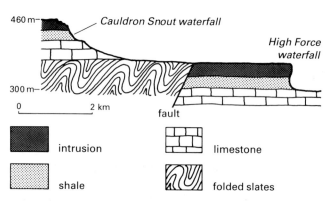

Fig. **1.28 (e)** Whin (?) along the Tees Valley, N.E. England

The History of the Earth

Fig. 1.29 will help you understand how old the Earth actually is.

35 The scale of Fig. 1.29 is 1 cm : 1000 years. How much longer will the line have to be drawn if you want to show the beginning of the Earth, 4600 million years ago? Identify a place that is that distance from where you are sitting now.

Processes and Forms in the Crust

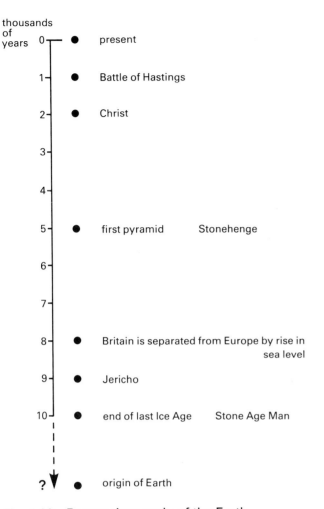

Fig. **1.29** Recent time scale of the Earth

to the youngest at the top. The separation into the different periods is based largely on fossil evidence within sedimentary rocks. The wavy line shows the times of great crustal movements or mountain building (**orogenesis**).

You will notice that in the upper part of the Carboniferous period, nearly 300 million years ago, reference is made to **Pangaea**. This is the name given to the single large continent that is thought to have existed at that time. We have seen earlier in this chapter that convection currents in the earth's mantle move the crustal plates in various directions. Evidence from rock strata, life forms, fossils and glacial erosion features is being collected to support the idea that Pangaea broke up into the continents we know today. This break-up may have been due to the convection currents and started about 180 million years ago. The continents are still moving as they 'float' on the crustal plates: this is the theory of **Continental Drift**.

37 a) Fig. 1.31 shows the positions that the continents may have occupied at various times. What new ocean has been created? What is the modern remnant of the Tethys Sea?
b) Describe what seems to have happened to India.

One remarkable fact is the very short time that civilised man has been in existence in relation to the life of the planet. The average lifespan of one individual would only measure about 0.75 mm on Fig. 1.29. It is worth bearing this in mind when studying the processes that create the surface features of the Earth. These have been in operation for thousands of millions of years. Any significant changes that take place in most landforms take a lot longer than the life of one geographer who studies them.

36 Which processes that you have studied in this chapter seem to create the most rapid changes?

Study Fig. 1.30. The column on the left is called the **geological column**. Its segments represent the sequence of rock formations from the oldest known at the bottom

Young fold mountains mark the boundaries between certain plates. Read again pages 5-8 and revise why this is so. From Fig. 1.30 you can see that mountain building has gone on periodically throughout the history of the Earth. It is not unreasonable to suppose that the earlier mountain chains shown in Fig. 1.2 represent the 'fossilised' positions of earlier plate boundaries. This emphasises that the present plates are not stationary and the present continental positions are probably not permanent.

The processes described in this chapter are dynamic, starting long before life appeared on our planet. Volcanicity, mountain building, earthquakes and continental drift are likely to continue for millions of years after mankind has disappeared.

38 Use an atlas to identify the **Hercynian** mountains G to J and the **Caledonian** mountains K to M on Fig. 1.2.

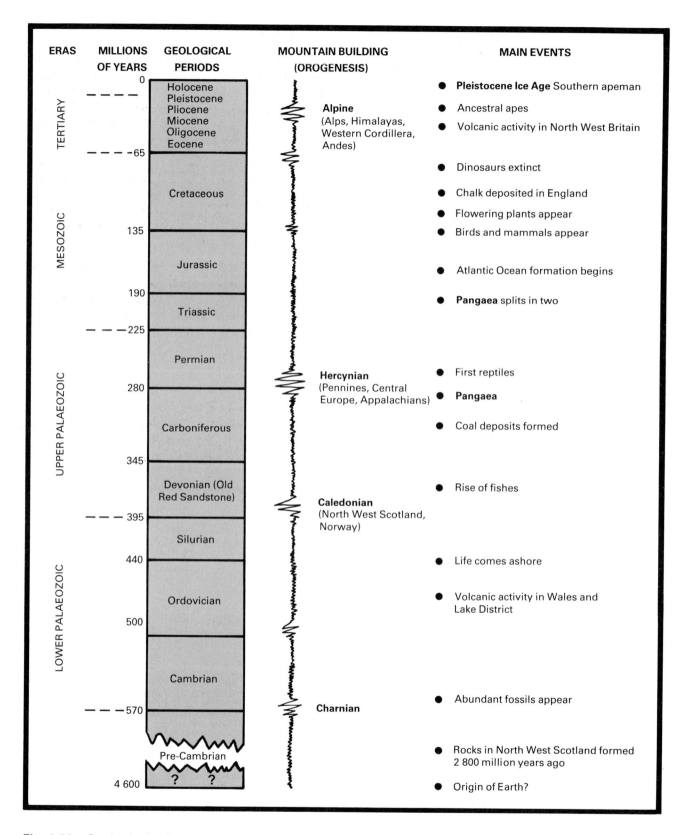

Fig. **1.30** Geological column and Earth history

Processes and Forms in the Crust

Workshop

1 From a map of the world, trace and cut round the outlines of the continents. How well can you fit them together? Trace them again, but this time use the outer edge of the continental shelf (200 metre submarine contour). Do they fit any better? Try to think of as many reasons as you can why a perfect fit is not possible.

2 Do a short project on two contrasting volcanoes. Include explanations of why they have been formed, their effects upon the people of the regions in which they are found, and an explanation of why they are different from one another. You should try to put in as many maps and illustrations as you can.

3 Start a collection of rock and mineral hand specimens. How might you set about classifying them? Each item must be carefully recorded, including its original location. You will find books on practical geology in your local library.

4 Explain carefully why the distribution patterns of volcanoes, earthquakes and fold mountains correlate so well.

5 With the help of a regional geography book of Europe, compare the landscape of the Hercynian mountains (e.g. the Harz, Hunsruck, Taunus and Schwarzwald) with that of the Alps.

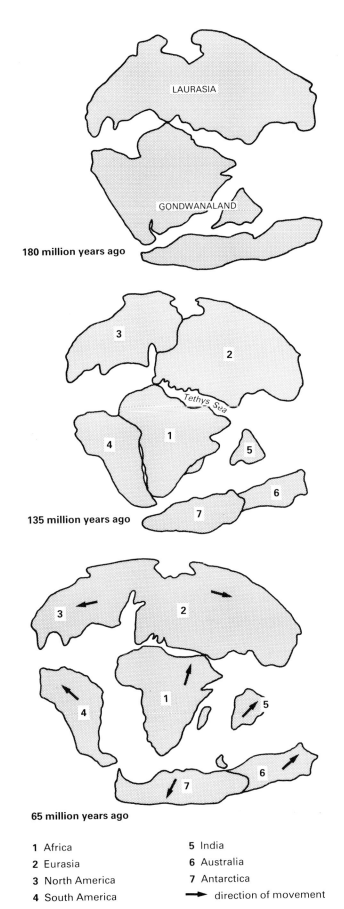

1 Africa
2 Eurasia
3 North America
4 South America
5 India
6 Australia
7 Antarctica
→ direction of movement

Fig. **1.31** Continental drift

Rocks and Relief

In the previous chapter we studied processes and landforms on a large, or macro-, scale. In this chapter we shall be looking at relief on the micro-scale: the local and regional level. Relief can be defined as the shape of the land. At this scale the factors that dominate landform development are **weathering**, **erosion** and **deposition** on the one hand, and the **structure** and **lithology** of the rocks on the other. Rock structure is the way in which strata have been laid down and distorted by folding and faulting. The lithology of the rock is the character of the rock itself: its chemical and physical properties.

It is important at this stage to be able to distinguish clearly between weathering and erosion. Weathering is the physical and chemical breakdown of the rock by the agents of the weather, chiefly rain, frost and temperature changes. Erosion is the wearing away of the rock, including the transportation of the material, by streams, ice, waves and wind. We shall be looking more closely at these agents of erosion in later chapters. **Denudation** includes weathering and erosion, and refers to the general lowering of the landscape.

Weathering

Mechanical weathering breaks the rock down into smaller particles which collect at the foot of slopes to form **screes**; these can be seen in Fig. 2.1.

Frost shattering, or **freeze-thaw**, is a common type of mechanical weathering. When the temperature falls below 0°C, water contained in cracks and crevices will freeze and expand, exerting great pressure on the rock, causing it to weaken. When the temperature rises and the ice melts, any particles loosened by the weakening will eventually fall away. The results of this can be seen locally in winter when holes develop in road surfaces and brick walls flake.

> **1** Study Fig. 2.1. Describe the relief of the mountain face. Draw a sketch of the scene and label the frost-shattered zone and the scree accumulations.

Fig. **2.1** Scree slopes at the foot of frost-shattered mountain sides

> **2** The requirements for freeze-thaw are fluctuations of temperature above and below 0°C and a supply of moisture. Use the climatic data in Chapter 4 to discover the regions of the world where this process is most likely to occur.

Exfoliation is due to wide temperature changes from very high (35°C or more) to relatively low (10°C) over a short period of time, such as day to night. Such a rapid and extreme change causes the rock surface to expand and contract with thin layers peeling off: for this reason it is also referred to as **onion-skin** or **spheroidal** weathering. Rocks in deserts are most prone to this, but it is also found in cooler regions, and it has been shown that some moisture has to be present for exfoliation to take place.

Fig. 2.2 Onion-skin weathering of basalt in Antrim

> **3** Fig. 2.2 shows the effects of exfoliation on a rock outcrop. Describe the results.

As the name implies, **chemical weathering** involves chemical changes in the minerals of the rock, in some cases dissolving them away completely. It occurs because rainwater is a weak acid, becoming even more acidic after being absorbed in the soil. The nature or lithology of the rock itself is very important in determining the type and effectiveness of chemical weathering. Cracks and fissures, known as **joints**, allow acidic water to percolate into the body of the rock, so deepening the effects of decomposition. Roots of plants and tunnels made by burrowing animals are also important in allowing chemical weathering to penetrate to lower levels.

Different minerals will react in different ways. Quartz in sandstones and granites remains unchanged; felspars in granitic rocks are changed into clay minerals and soluble salts; some limestones are dissolved (see p.37).

> **4** If you have access to a collection of rock specimens, test each specimen with a drop of dilute hydrochloric acid. How does each type react?

One important result of weathering, both mechanical and chemical, is the accumulation of material known as the **weathered mantle** or **regolith** on top of the rock itself (Fig. 2.3). In the upper part this is mixed in with organic material to form the **soil**.

> **5** Study Fig. 2.1.
> a) How much material has accumulated on the *upper* slopes of the mountainside? Why do you think this may be?
> b) Estimate the angle of slope of the bare rock face and the angle of the scree slope beneath it.
> c) On what kind of slope would you expect to see the thickest build-up of weathered material?

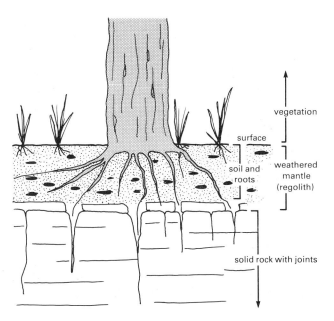

Fig. 2.3 The weathered mantle

You should now be able to see that the angle of slope and the effects of gravity are very important factors in the amount of weathered mantle that can build up. In addition, the type of weathered material itself will influence the angle of slope. Large particles accumulate with steeper slopes than fine particles.

> **6** Turn to the first O.S. map extract on page 127.
> **a)** Measure the gradient from the spot height at 568836 to the spot height at 568843.
> **b)** Where is the steepest slope on the extract? Measure its gradient. By referring to the geology map (Fig. 2.4), find the kind of rock in which this slope has been formed.
> **c)** Measure the gradient between Lollingdon Farm (571854) and Westfield Farm (578856). In what kind of rock material is this slope formed?

Slopes

When you look at a landscape, either in real life or from a map or photograph, it is noticeable that *slopes* are a dominant feature. The study of slopes, their formation, gradient and shape, has become a very important part of modern physical geography.

Clay is much softer than chalk and cannot support steep slopes. But this does not mean that you do not find gentle slopes on chalk or other harder rocks: look for example at the area just to the north-east of Breach House (579844). No doubt you can find other flat areas on the chalklands of this extract.

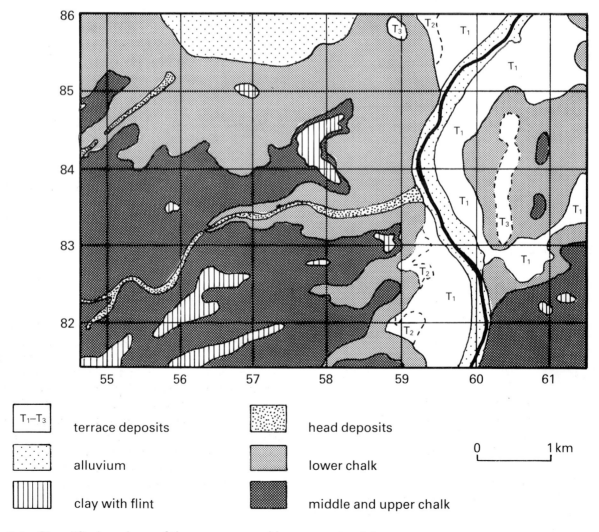

Fig. **2.4** Simplified geology of the area covered by map extract 1

Rocks and Relief

The shape of a slope, its **profile**, is as important as its gradient. Fig. 2.5 illustrates four basic profile types: straight, concave, convex and concavo-convex.

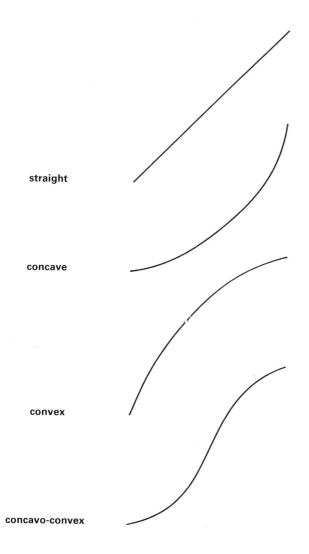

Fig. 2.5 Four simple slope profiles

In fact, concavo-convex slopes appear to be very common in chalk country, but in order to test this thoroughly you would have to take many more sample profiles than just one.

The second O.S. map extract shows an area of old, hard sedimentary rocks just to the north of Aberystwyth in Wales. These rocks are much harder and more resistant than the chalk around Goring on map extract 1, and there is also another very important difference.

> **9** Study both O.S. map extracts. What contrast can you see in the amount of surface water (rivers and streams) on the two areas?

The chalk is **permeable**, which means that water passes through it, hindering the formation of streams on the surface. The rocks in the second area are **impermeable**, resulting in many surface streams. We might expect to find that hard, impermeable rocks will have slopes that are different in profile and gradient from rocks that are less hard and permeable.

> **10 a)** From O.S. map extract 2, draw the profile of the slope from the trig. point (triangulation pillar) at 666867 to the stream at 666859. What kind of profile have you drawn? Measure the gradient of the slope.
> **b)** How does this slope compare with the slope you studied in exercise 8?

> **7** Refer to Fig. 2.1. Which one of the four profile types do you consider the scree slopes to be?
>
> **8** Turn to O.S. map extract 1.
> **a)** Draw an accurate profile from the tumulus at 579828 to the road at 579834. Make sure you use a sharp pencil as the contours are closely spaced.
> **b** What kind of slope have you drawn? Describe the spacing of the contours on the map for such a slope. Now measure the average gradient of this slope.

It would be far too hasty to jump to any conclusions about the differences between the slopes of the two areas on the basis of only two sample slopes (and carefully chosen ones at that!). In order to test the hypothesis *that the slopes in the Goring area are less steep and of different profile in comparison with the slopes of the Aberystwyth area*, we must study a larger number of samples. We still have to work with samples because it would be a very difficult job to measure all the slopes in each area. For our samples we are going to take ten grid squares at random from each O.S. map extract.

11 a) *For the Aberystwyth extract*: restrict your survey to the area between eastings 61 and 66 and between northings 84 and 87. Now proceed as follows:

Consult the table of random numbers on page 29. You will notice that all the eastings on the survey area we have chosen start with the digit 6, and all the northings start with the digit 8. To select a square at random, take a digit from the random number tables. Put 6 in front of it. Take the next number from the tables. Put 8 in front of it. For example, say the figures you now end up with are 65 and 85. This then is the south-west corner of square 6585. Some numbers you pick will give you a square that is outside the survey area: these will have to be ignored.

b) *For the Goring extract*: restrict your survey to the area between eastings 56 and 61 and between northings 82 and 85. This will give you the same size of survey area as for the Aberystwyth extract. Change the numbers of eastings 60 and 61 to 50 and 51. Select ten squares at random but this time all eastings will start with 5 and all northings with 8.

In selecting digits from random number tables, you can start anywhere in the tables, but then all digits must be taken in a regular order either along the rows or down the columns. Although the digits are arranged in pairs, you can take them singly.

c) When you have selected your squares, take the measurements listed below for each one. Record your results in a table for each extract. Table 2.2 shows the results for just one square on the Goring extract.
 i Find the highest and lowest points, in metres.
 ii Measure the distance between them, in metres.
 iii Work out the gradient (see page 116).
 iv Note the shape of the slope profile from the contour pattern.
 v For each extract work out the mean (average) gradient from the ten samples.

Note that the two extracts are not of the same scale and do not have the same contour interval.

d) Is there much difference between the mean gradients for each extract? Have these results confirmed our hypothesis?

Any difference that you find between the two means might in fact be due to **chance** because we have only taken ten sample squares from each extract. Many more samples would be necessary to show that there is a **significant difference** between the two areas, i.e. that they really do have different mean gradients. In Chapter 7 we will investigate a method of finding out whether or not differences between measurements are significant or not (see page 176).

Slope Systems

There are many factors in addition to rock type that will influence slope gradient and profile. Fig. 2.6 summarises the main ones. All the factors in the diagram are working together to produce a slope. This working together of a variety of factors to produce a particular result is known as a **system**. Fig. 2.6 is a simple slope system, and you will notice that this system has **inputs** and **outputs**. Inputs are factors that come into the system from outside the system itself, and outputs are factors that leave the system.

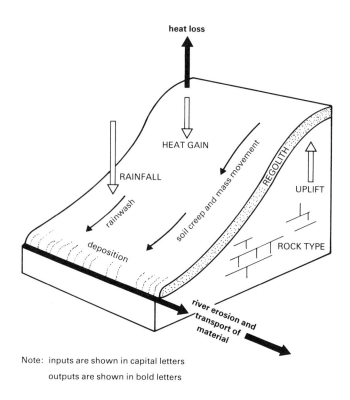

Note: inputs are shown in capital letters
outputs are shown in bold letters

Fig. **2.6** A simple slope system

12 Study Fig. 2.6 and list the inputs and the outputs.

Rocks and Relief

04 57	45 89	01 69	00 66	98 33	02 86	73 01	27 97	92 17	32 61
60 96	96 13	48 83	99 24	24 06	39 51	70 45	42 09	57 83	78 81
02 35	50 67	01 26	49 34	32 63	55 62	09 92	21 68	95 66	22 05
45 78	15 35	28 44	35 96	34 43	86 98	39 52	83 40	69 08	74 63
26 23	88 15	02 47	76 49	73 09	95 88	90 70	29 13	77 20	32 69
60 24	55 37	92 54	13 71	68 10	31 19	84 22	15 09	37 93	74 79
68 36	43 45	39 61	43 67	91 60	29 18	18 97	29 98	42 25	81 19
38 18	07 10	29 59	46 70	75 50	15 72	84 38	00 33	08 50	35 98
78 99	70 77	92 56	50 08	67 55	83 93	82 29	50 02	93 16	11 61
46 69	39 94	24 15	14 79	78 63	46 36	38 42	47 07	09 96	11 03
61 38	06 56	28 60	11 69	29 57	14 05	84 30	11 24	98 63	45 28
29 09	18 02	84 80	29 78	55 81	33 40	94 37	57 46	06 43	67 72
85 62	94 11	01 54	42 11	25 51	66 97	06 87	81 93	86 36	95 17
86 31	37 26	68 64	03 08	32 77	34 75	31 04	78 06	58 03	47 22
18 15	03 46	57 08	37 11	67 18	98 06	99 11	20 09	93 37	60 41

Table **2.1** Random numbers

No.	Grid Sq.	Highest point	Lowest point	Distance	Gradient	Profile
1	5684	108	62	1200	1 in 26	concavo-convex
2						
3						
.						
.						
10						

Table 2.2 Slope readings for the Goring extract

We have already seen how rock type can influence slopes. We shall now consider some of the other inputs. **Crustal movements** may result in **uplift** of the surface. This will influence the height of the land and therefore how steep the slopes can be. The steeper slopes in the Aberystwyth area are in part due to the land there being higher in the first place.

> **13** From the data obtained in exercise 11, work out the average height of the Aberystwyth area and the Goring area.

Energy in the form of **heat** from the Sun is another important input, but it is also an output: we have seen how temperature changes cause freeze-thaw and exfoliation, resulting in rock weathering and the formation of the regolith. In the next chapter we shall see how heat is important in the formation of **rain** which is a further important input into the system. Raindrops can fall with a considerable impact on the ground, breaking up soil particles and washing them down the slope (**rainwash**). Furthermore, the rainwater lubricates the regolith material, helping it to 'creep' downhill to be deposited at the bottom of the slope. Remember also that acidic rainwater actually helps to create the regolith in the first place. Farmers can suffer severe loss of soil because of **soil creep**, the movement of the soil down the slope. Ploughing a field parallel to the contours helps to prevent this happening.

On steep slopes, movement of material can be on a dramatic scale, resulting in **mass movement** which includes landslides (or landslips) and avalanches, especially after heavy rain. In California a landslide involving 1800 cubic metres of material slipped down the Tin Mountain, covering an area of 14 square kilometres. Fig. 2.7 illustrates some more modest slips on the Isle of Wight. When the marls become waterlogged in wet weather they are unable to support very much weight, including their own, causing them to collapse and slip downhill. Folkestone Warren in Kent is another place famous for its landslips.

> **14** Study Fig. 2.7 and describe the profile of the present slope. Attempt a reconstruction of the original slope before any slips took place.

Notice that the movement of the slipped masses in Fig. 2.7 has been in a curved motion called a **rotational slip**.

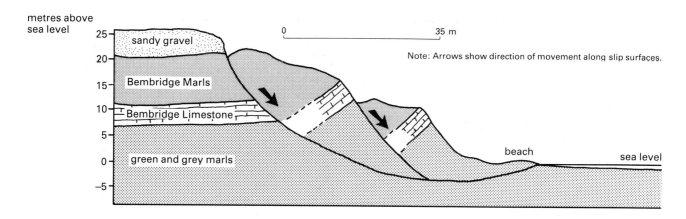

Fig. 2.7 Landslips on the Isle of Wight

Soil creep, mass movement, rainwash and deposition are processes *within* the system and are not regarded as inputs or outputs. Amongst the most important processes are the erosion and transportation (i.e. output) of the material that has been deposited at the base of the slope. If the stream did not remove this material, the deposits would get deeper and deeper and start to accumulate uphill. If the stream removed more material than was being deposited, it would start to erode the rock itself at the foot of the slope along the valley side.

> **15** How might a slope *profile* change if:
> a) material was not removed by a stream from the base of the slope, and
> b) more material was removed than could accumulate at the base of the slope?

It may be that the difference in drainage between the Goring area and the Aberystwyth area might help to explain the differences in slope profiles.

Cuestas and Vales

> **16** From O.S. map extract 1, draw an accurate profile from 560834 to 568845.

The slope that you have just drawn is a very important feature in this area and is called a **scarp slope**; it trends its way across the area as a steep hillside, almost like a cliff.

> **17** From the same map extract, trace off the path of the 100 m contour from 560840 to 570815.

This contour is about half-way up the face of the scarp. Notice how it sweeps around three prominent ridges. If you now join up the eastern ends of these ridges with a line, you will have produced the trend line of the scarp, showing the direction in which it runs. The scarp has been deeply eroded into, or **dissected**, creating the ridges separated by steep-sided valleys.

If we now look at the area to the east of the Thames, we can follow another scarp which runs north-east from Spring Farm at 605823. If we extended the map extract north-eastwards, we would see that the scarp on this side of the Thames is the scarp face of the Chiltern Hills.

> **18** In your atlas turn to a map showing the relief of southern England. Identify the Chilterns, the Thames and the position of Goring. To which hills does the scarp that we identified on the western side of the Thames belong?

Fig. 2.8 summarises the position of the scarp in the Goring area. It is thought that the scarp faces of the Chilterns and the Berkshire Downs were once one continuous scarp, but the River Thames has cut a **gap** through it known as the Goring Gap. Many similar gaps occur in the scarplands of southern England.

Rocks and Relief

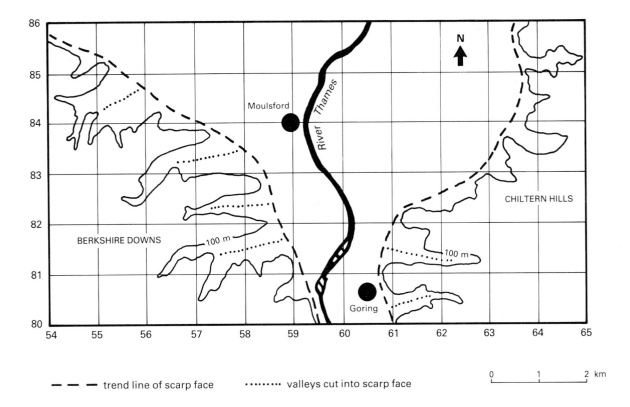

Fig. 2.8 The Goring Gap

19 Maidstone, Dorking, Guildford, Arundel, Luton and Goring are all towns that command gaps in chalk scarps. Using your atlas, draw a sketch map to show the positions of these towns; mark and name the rivers and shade areas over 150 m, naming the hills. How important are these gaps for transport routes to and from London?

Fig. 2.9 is a photograph taken at 573837 (Goring extract), looking south-east.

20 Draw a sketch of the view and on this label Greenlands Farm, the A 417 road, the site of a tumulus and the gap cut by the Thames.

21 a) Describe the shape of the ridge. What is its height?

b) How does the height of this ridge compare with the heights of the other ridges on the extract?

Fig. 2.9 View looking S.E. from 573837 on map extract 1

Your answers to the last question should reveal that the ridges are all fairly uniform in height, giving a plateau-like surface at the top of the scarp, which can be seen to its best advantage at 559836. This gently sloping surface is known as the **back slope** or **dip slope**, and, like the scarp itself, has been deeply dissected. Fig. 2.10 shows how scarps and dip slopes are formed from strata that dip uniformly into the crust.

You will notice that the vale does not lie right at the foot of the scarp which we have already identified (exercise 17). This is because there is a secondary scarp at a lower level than the main one. It can be seen in the contour lines that run east–west between Carrimers Farm (559849) and Lollingdon Farm (571854). Fig. 2.11 shows a diagrammatic cross-section along easting 56 between northings 83 and 86.

22 a) In what kinds of rock are the scarps and dip slopes formed? Why do you think that the dip slopes are so named?
b) What has caused the vales to form in between? (See page 101.)

24 With reference to Fig. 2.11 and the geology map on page 26, identify the rock types in which the main scarp is formed and in which the lower scarp is formed.

Together the scarp and dip slope are called a **cuesta**. The gaps in the cuestas are cut by consequent streams like the Thames at Goring. The vale in the Goring district can be seen in the far north of O.S. map extract 1.

23 With reference to the geology map on page 26, in what type of rock has the vale been formed?

Uniformly dipping strata of alternating bands of hard and soft rock will produce cuestas and vales. If the strata were of the same lithology, such a relief pattern would not be formed. If folding occurs with alternate hard and soft rocks, a different pattern again emerges. On page 9 we saw that the Jura Mountains have ridges on the crests of the anticlines and valleys in the troughs of the synclines. In many cases, however, the rocks on the crests get stretched, they crack open, become weaker and so are more prone to erosion. The rocks in the troughs on the other hand may become more resistant to erosion because they become compressed.

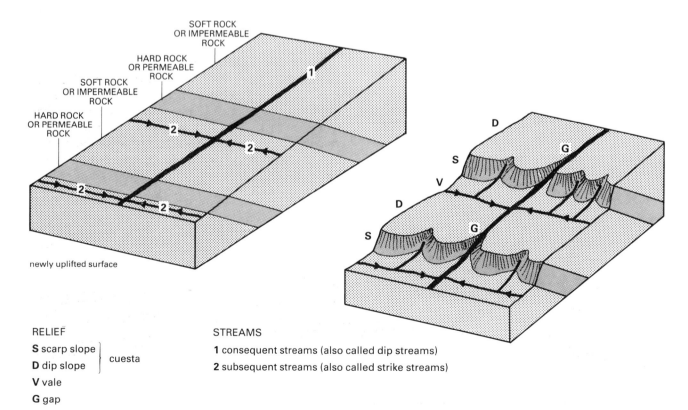

RELIEF
S scarp slope ⎫
D dip slope ⎬ cuesta
V vale
G gap

STREAMS
1 consequent streams (also called dip streams)
2 subsequent streams (also called strike streams)

Fig. **2.10** The formation of cuestas and vales

Rocks and Relief

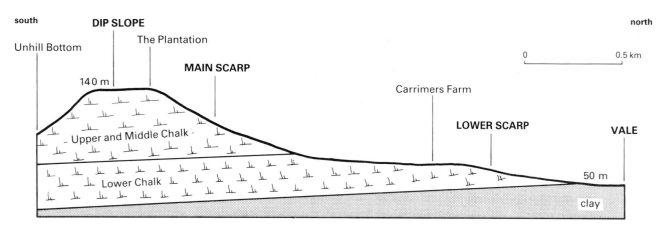

Fig. 2.11 Sketch section along easting 56, map extract 1

25 Study Fig. 2.12 which shows the weakening of rock on the crest of an anticline. Copy this into your exercise book and then draw a sequel to show the probable surface relief after erosion has taken place on the top layer. You should be able to label two scarps and two dip slopes.

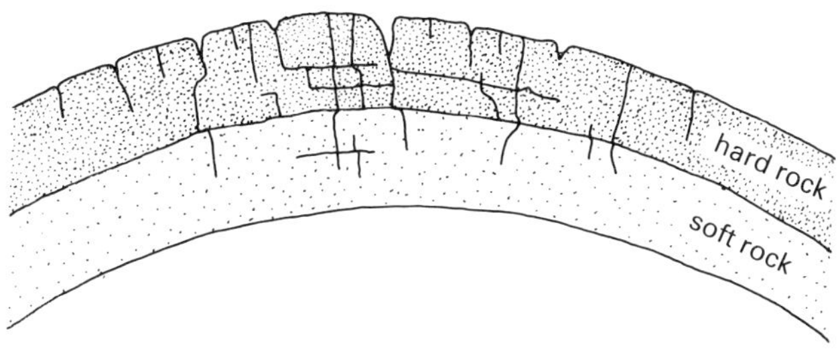

In fact, far from the anticline forming a ridge, it can form a vale. This has happened in the Vale of Pewsey, Wiltshire (Fig. 2.13) and is an example of what is known as **inverted relief**. Another example of inverted relief is shown in Fig. 2.14, a section through Tortworth, near Bristol.

26 Explain why Fig. 2.14 is an example of inverted relief. How does it differ from the example of the Vale of Pewsey?

Fig. 2.12 Rock weakening at the crest of an anticline

Anticlines can therefore form inward facing scarps across a vale, while synclines can form outward facing scarps, but only if the lithologies of the alternating rocks are different. Fig. 2.15 is a very good example of the effects of structure and lithology on relief. The simplified section runs from Cheltenham in the northwest, through Oxford to central London, and then turns due south to Brighton.

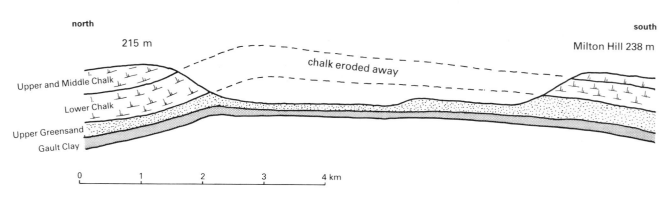

Fig. 2.13 Inverted relief: sketch section through the Vale of Pewsey

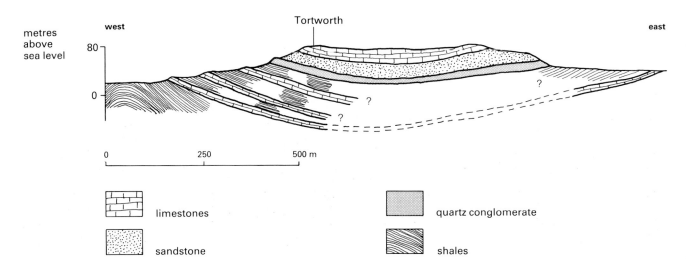

Fig. **2.14** Inverted relief: sketch section through Tortworth, near Bristol

27 Make a large copy of Fig. 2.15. With reference to the geology and relief maps in your atlas, label your diagram with the following hills: the Cotswolds, North Downs, South Downs, Hampstead Heath and the Chilterns. Mark the scarps with an 'S' and the dip slopes with a 'D'. Label the towns of Cheltenham, Oxford, Watford, Reigate, Crawley and Brighton. In the appropriate spaces below the section write 'The London Basin' and 'The Weald'. Give an approximate horizontal and vertical scale.

The Tertiary sands and clays that now occupy most of the London Basin were deposited by a sea that flooded the area between the Downs and the Chilterns some 60 million years ago.

Faulting is another kind of structure that can produce scarps, as illustrated in Fig. 2.16. Cross Fell has been uplifted to give a **fault-scarp** overlooking the Vale of Eden. This fault-scarp is on the western edge of the tilted plateau of the North Pennines. Denudation after faulting sometimes removes the uplifted scarp face altogether, especially if the rock is weak; but in the case of Cross Fell, Carboniferous Limestone is very resistant and so the scarp has been preserved.

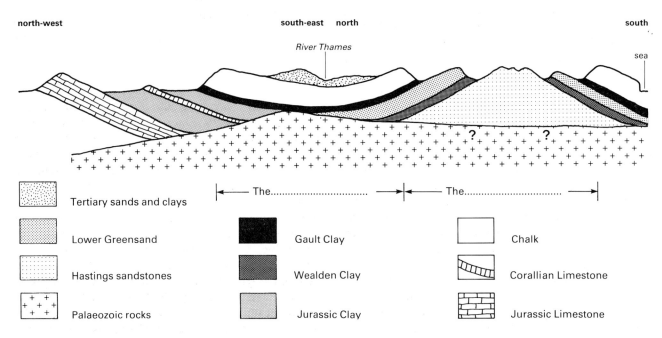

Fig. **2.15** Sketch section through lowland England

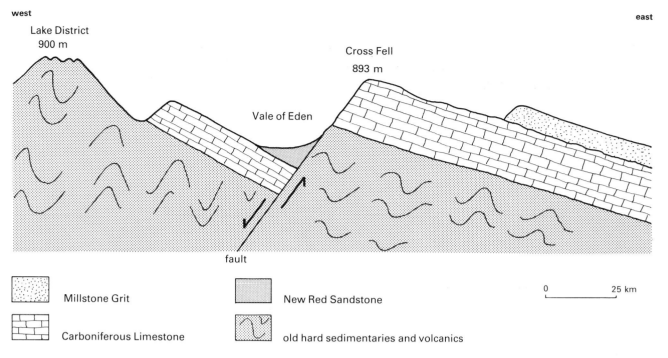

Fig. 2.16 Sketch section through the North Pennines

Rock Types and Relief

We have noted that cuestas would not be developed if all rocks were of the same lithology. Certain rocks do create quite distinct relief features because of the way in which they respond to denudation processes.

Chalk Relief

A good deal has already been discovered about the chalk relief depicted on O.S. map extract 1. One important characteristic of these chalklands is the general absence of surface water, except for the Thames. Drainage channels can however be found in the north of the map area where alluvium overlies clay. Clay is an impermeable rock, that is, it does not allow water to pass through it. The grains out of which clay is made are extremely small: therefore the pore spaces between them are extremely small. Water can be held in the pore spaces by capillary action and this has the effect of preventing surface water from **percolating** downwards. The grains out of which chalk is made are larger, making the pore spaces between them large enough for water to flow through. In addition, the rock is fretted with fine, irregular cracks which increase the rock's permeability.

However, despite the rock being permeable, chalklands are covered with a network of valleys. These can clearly be seen on the Goring sheet, an example being that in square 5782. Because they do not contain streams, they are known as **dry valleys**.

28 Place a piece of tracing paper over the Goring sheet. Trace off the pattern made by the dry valleys by using a pecked (dashed) line along the valley bottoms. Describe the pattern made by your lines. Into which valley do most of them converge?

The pattern you have just described shows all the features of a stream network, but without streams. If there are no streams, how could the valleys have been formed? Or, putting the problem another way, where did the water go after it had presumably eroded the valleys? It is necessary to look more closely at the drainage of chalk country. Fig. 2.17 shows the distribution of water within a chalk cuesta. Rain falling on the chalk percolates through the rock because it is permeable. As the water cannot flow through the impermeable clay that underlies the chalk, it builds up in the chalk (**the zone of saturation**), the upper surface of the saturated zone being called the **water table**. At the foot of the scarp where the clay comes to the surface, the water table emerges as a series of **springs** along a **spring-line**. The foot of the dip slope also has a spring-line.

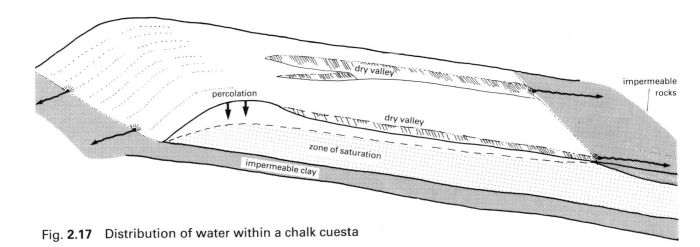

Fig. 2.17 Distribution of water within a chalk cuesta

29 Make a copy of Fig. 2.17 and label clearly the following: scarp slope, dip slope, springs, water table.

30 Give the six figure grid reference for the spring marked in the north-west of O.S. map extract 1. Although it is not labelled, there is a spring to the east of Lollingdon Farm. Give its six figure grid reference. Is it at the same height as the first spring?

31 Referring to Fig. 2.17, what do you notice about the relative heights of the water table and the bottom of the dry valley?

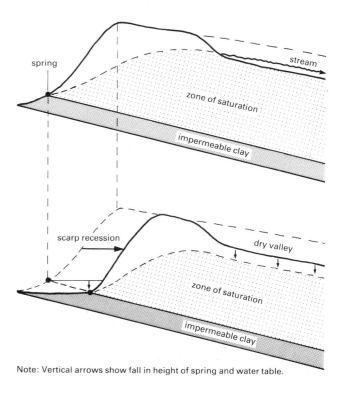

Note: Vertical arrows show fall in height of spring and water table.

Fig. 2.18 A possible mode of origin of chalkland dry valleys

This difference is the reason for the dryness of the valleys, for if the water table were higher it would intersect the valley bottom, and a stream would flow. In wet weather the water table does in fact rise, and in some areas it rises high enough for dry valleys to become wet. Such wet weather streams are known as **bournes**. So, for the valleys to have been cut in the first place, we can assume that the water table was once permanently higher than it is now. Why has it fallen?

One suggestion is that the scarp face is receding because of denudation. As the base of the chalk – its junction with the clay – dips downwards, so the spring-line falls in height, creating a corresponding fall in the height of the water table. This is shown in Fig. 2.18. This theory has not been proved, however, and there are other ideas, including the periglacial hypothesis outlined in Chapter 6 (on page 158). Although dry valleys are an important characteristic of chalklands, they also occur in other limestone types and in permeable sandstones.

Permeable rocks like chalk provide considerable reserves of water for domestic and industrial uses. Dig a well down to the water table and the water can be pumped up to the surface. If the top of the well is below the highest point of the water table, the water will rise to the surface under its own hydrostatic pressure (an **artesian well**). Fig. 2.19 is a simplified cross-section through the **aquifer** (water-bearing rock) of the London Basin.

32 Look at Fig. 2.19. What do you notice about the difference in height between the top of the first well (W_1) and the water table, and the top of the second well (W_2) and the water table? Which of the two wells will need pumping, and which will yield water naturally?

There is a natural flow of water from the chalk into the Thames, but because London is extracting so much water from the aquifer, the flow is being reversed in the east of the basin, causing polluted Thames water to flow back into the chalk. With a population of over 10 million people, each person using, on average, 275 litres of water a day, you can see that the demands on the natural supplies are very great. The aquifer itself cannot supply all this water. A large amount comes from the Thames and its tributaries, and there are plans to divert water from the Severn Basin into South East England.

33 Springs provide valuable local water supplies for farms sited along the bottom of the scarp and dip slopes of permeable cuestas. Can you suggest any other advantages of such a location for farms?

The main features of chalk relief can be listed as follows: smooth concavo-convex slopes, dry valleys, scarp slopes and dip slopes. However, it must be borne in mind that other rocks, such as the Jurassic Limestone of the Cotswolds, will also have these features. When trying to identify chalk from a strange map extract, place-name evidence does help a lot. The occurrence of names like 'Down' (as in Thurle Down, square 5781), 'Bottom' (as in Moulsford Bottom, square 5883) and 'Coombe' (not represented on this extract) often indicate chalk. Neolithic peoples tended to settle on the chalk uplands of southern England, one reason being that the thin woodland and light soils made agriculture easier there than on the heavily wooded clays in the vales.

34 What evidence of Neolithic settlement is there on the Goring sheet?

Limestone Relief

Chalk is a pure form of limestone (calcium carbonate) but it is soft in comparison with some other limestones which can be very hard and crystalline. These types include the **Carboniferous Limestone**.

35 Turn to a geology map of Britain in your atlas and locate the areas of Carboniferous Limestone.

Like all limestones, Carboniferous Limestone is soluble in water, particularly acidic water. Rainwater becomes a weak carbonic acid due to the presence of carbon dioxide in the atmosphere, in the soil and in the rock itself. The chemical reaction can be written like this:

$$CaCO_3 + CO_2 + H_2O \rightleftharpoons Ca(HCO_3)_2$$

calcium carbonate + carbon dioxide + water
$\rightleftharpoons$ calcium bicarbonate

The calcium carbonate is removed in solution as calcium bicarbonate.

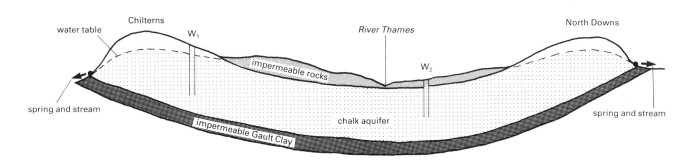

Fig. **2.19** London's artesian basin

Carboniferous Limestone is massively jointed, that is to say it has a well defined pattern of deep cracks, both horizontal and vertical. It is these joints that make the rock permeable and it is along these joints that the solution process is concentrated. It is the lithology of Carboniferous Limestone – its hardness, solubility and jointing system – that gives rise to distinctive landforms. Figs. 2.20, 2.21 and 2.22 illustrate the main relief features near Malham, Yorkshire.

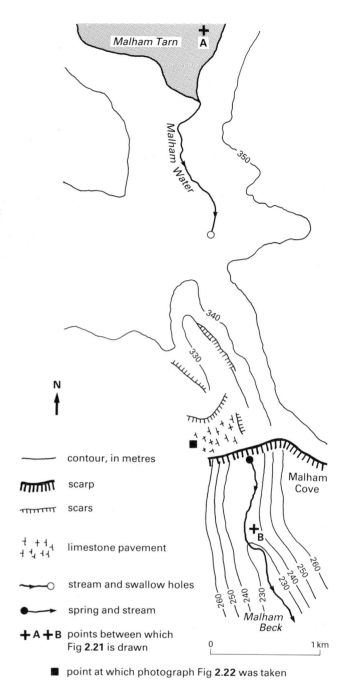

Fig. 2.20 Sketch map of the Malham area

> **36** Study Figs. 2.20 and 2.21.
> **a)** Why does Malham Water disappear about 1 km south of Malham Tarn? Your answer should refer to the structural features shown as well as to the lithology of the rocks.
> **b)** Why should a stream emerge from the base of the scarp at Malham Cove?

Do not assume that the water disappearing just to the south of Malham Tarn is the same as that which appears at the foot of the cove. Research shows that it is not! Malham Water comes to the surface further south, while the water in Malham Beck comes from swallow holes further to the west than those shown on Fig. 2.20.

> **37** What feature is shown by the contour lines on Fig. 2.20 between the swallow hole and the top of the scarp? What does this suggest about the possible movement of the water table?

Fig. 2.22 is a photograph taken from the point marked with a square on Fig. 2.20, looking south-east.

> **38** Describe the nature of the surface which can be seen in the foreground.

Such a surface is called a **limestone pavement**; the main features of it are identified in Fig. 2.23.

> **39** How do you think such features were formed?

Grykes may be deepened by the acids produced by the scanty vegetation that often grows in the bottom of them. A **scar** is a very steep face of a particularly hard band of rock that has been exposed by denudation.

Swallow holes are small versions of the many enclosed holes and depressions that occur all over the surface of limestone regions. Larger ones are called **dolines**, measuring several metres across and several metres in depth. **Poljes** are very much larger than this, sometimes several kilometres across. Such large hollows may be caused by surface solution (the action of rainwater

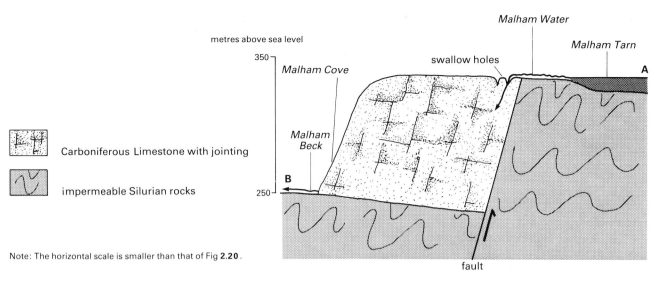

Fig. 2.21 Sketch section from Malham Cove to Malham Tarn

described earlier) like their smaller relations, with several dolines merging together. Alternatively, they may be due to the collapse of cave systems below the ground, resulting in subsidence on the surface. Poljes do not feature in British limestone regions but they are found in North West Yugoslavia in an area called the Karst. In this region, limestone features are very well developed. Thus the region has given its name to limestone features in general: they are termed **karst** features.

Areas of karst, which include the Massif Central of France, have poor, thin soils. The poljes, however, are floored with clay, a residue from weathering, and they provide places for more productive agriculture.

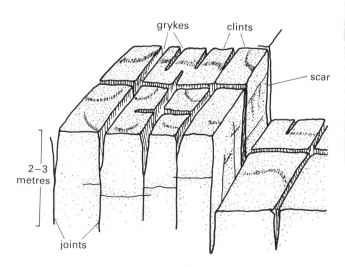

Fig. 2.23 Limestone pavement

Fig. 2.22 Limestone pavement above Malham Cove

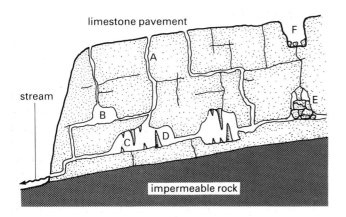

Fig. 2.24 Underground features in Carboniferous Limestone

Underground solution takes place to produce pot holes, caves, caverns and underground stream channels, shown in Fig. 2.24. Inside the caves, droplets of percolating calcium bicarbonate solution evaporate on the ceiling, causing the accumulation of calcium carbonate in a downward direction to form **stalactites**. Some of the solution drips to the floor to create **stalagmites**. Caverns may be linked to the surface by **pot holes**, vertical or near vertical shafts hollowed out by solution. The caves and caverns are joined together by underground stream channels. The smaller caves and channels can become filled with torrents of water after heavy rain on the surface: caves which are beneath the water table are permanently flooded. Cavern collapse creates **gorges** at the surface, a famous example being Cheddar Gorge in the Mendips.

40 Make a large copy of Fig. 2.24. Provide a key for all the features labelled A to F. Assess the importance of jointing in the rock in the formation of these features.

The karst features described above are found in the Carboniferous Limestone areas of Britain. The other major outcrops of limestone, chiefly the Jurassic Limestone of the Cotswolds and Northamptonshire Uplands, do not develop such features. Although permeable and soluble, they are too soft and lack a well defined joint system. However, the Jurassic Limestone in central France and N.W. Yugoslavia does have the right properties for karst development. It is the lithology of the rock, not its age, which is important.

Igneous Relief

The major relief features to be found in areas of igneous rock have been dealt with in Chapter 1 (pages 13–20). Locally, igneous rocks give rise to distinctive landforms where the rock has contracted on cooling to produce a well defined pattern of joints. We have already seen the effects of columnar jointing in basalt at the Giant's Causeway (page 14).

Granite develops a massive jointing system. This has given rise to the famous **tors** on Dartmoor and neighbouring granite areas like Bodmin Moor and Hensbarrow. These areas are igneous intrusions that have been exposed by the denudation of the rocks above them. Fig. 2.25 is a photograph of Hay Tor on Dartmoor.

Fig. 2.25 Hay Tor, Dartmoor

41 Study Fig. 2.25. In what directions have the joints developed? What influence has the jointing system had on the shape of the tor?

The way in which tors are formed has been a matter of some debate among geographers. Fig. 2.26 summarises one idea.

42 Study Fig. 2.26.
a) What happened to the granite beneath the original ground surface?
b) What is the relationship between this process and the density of jointing in the granite shown in parts A and B of the diagram?
c) What process established stage C?

43 In Fig. 2.25 estimate the height of the original ground surface above the present ground level.

Rocks and Relief

Another idea is that the tors were formed during the last Ice Age. Although Dartmoor was not covered by an ice sheet it was under the influence of periglacial conditions (see pages 157–8). Freeze-thaw processes would have weathered the rock deeply along the joints, and solifluction (page 158) would have removed the debris. Tors are not confined to granite. For example, in central Otago, New Zealand, they have been formed in metamorphic schists. Brimham Rocks near Harrogate in Yorkshire are tor-like features in Millstone Grit, although the mode of formation may not be the same as that of the Dartmoor tors.

Workshop

1 With reference to Fig. 2.15 describe how the structure and the lithology of the rocks has influenced the relief of lowland England.

2 Construct a simple clinometer (an instrument for measuring slopes) from a wooden ruler, protractor and plumb-line, as shown in Fig. 2.27. A partner of your own height should stand a measured distance away. In the example given in Figs. 2.28 and 2.29 this measured length is 20 m. Look along the ruler and sight onto your partner's eyes. A third person can read the slope angle from the point at which the cord intersects the protractor. The result is the angle of slope of the ground along the measured length.

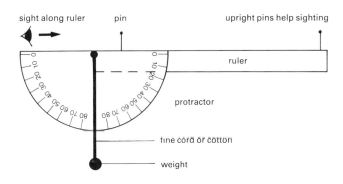

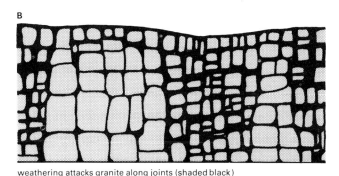

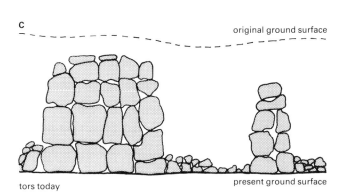

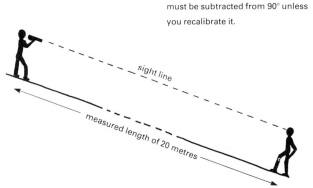

Fig. **2.27** Construction and use of a simple clinometer

Fig. **2.26** A possible mode of origin of tors

Select a transect, or section, in your local area: you might choose a scarp face or a small valley. Try to get your transect to go up and down the maximum slope angle, that is, the path that would be taken by a freely rolling ball. Do keep off private land unless you have obtained permission to cross it beforehand, and under no circumstances trample down crops. Do not make your transect too long: remember that you will have to make 50 readings in one kilometre if the measured length is 20 m!

Record the results as shown in Fig. 2.28, and map them as in Fig. 2.29. How might you set about constructing a profile drawing of your slope?

SLOPE READINGS							
START COLNE FARM 826520				FINISH DOG & POT P.H. 828520			
MEASURED LENGTH = 20 metres							
measured length	angle (degrees)	up-hill	down-hill	measured length	angle (degrees)	up-hill	down-hill
No: 1	6		✓	6	1	✓	
2	5		✓	7	2	✓	
3	4		✓	8	4	✓	
4	2		✓	9	7	✓	
5	1		✓	10	9	✓	

Fig. **2.28** Record of slope measurements

Is your slope uniform, or does it have distinct breaks (or changes) of slope? If it does, can these be explained by changes in geology? Are breaks of slope marked by changes in land use or vegetation? Is there a relationship between soil depth and slope? For this problem you would need a soil auger (a sort of giant corkscrew) to bore into the soil to measure its depth.

There is quite a lot of scope for hypothesis testing project work on slopes, but remember that you cannot prove anything from just one transect, and as with some of the work done with the O.S. map extracts in this chapter, you will need a number of samples. If you live in an urban area there is scope for slope analysis in the local park, but it can be done in the street too. One hypothesis that could be tested in a town is *that type of housing differs according to the slope and height of the land.*

A Note on Hypothesis Testing and the Scientific Method of Enquiry

It has been suggested that you attempt to test one or more **hypotheses** about the relationships between slopes and other factors like soil depth and vegetation. Put simply, a hypothesis is an idea, or a hunch, that you have about something you are studying. It should be made up in the form of a relationship between two factors, or **variables**. For example, you may come up with this hypothesis: *that as a slope gets steeper, so soil depth decreases.* The two variables in this case are the slope angle and the soil depth. They are called variables because they vary, or change, in quantity.

Having established your hypothesis, it needs to be tested by practical work or fieldwork. Analysis of the results can lead you to accept or reject your hypothesis. If accepted, then the ideas in the hypothesis can contribute towards the general body of theory about the subject. If rejected, you may have to rethink and change your original hypothesis. This sequence of events is summarised in Fig. 2.30 and is known generally as the **scientific method of enquiry**, now used widely by geographers in physical as well as human geography.

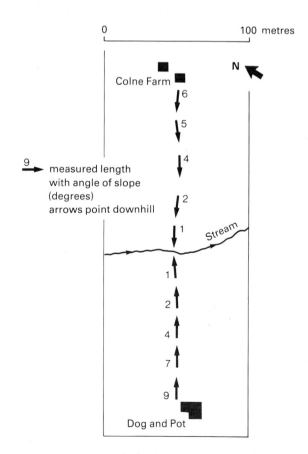

Fig. **2.29** Map of slope readings along transect

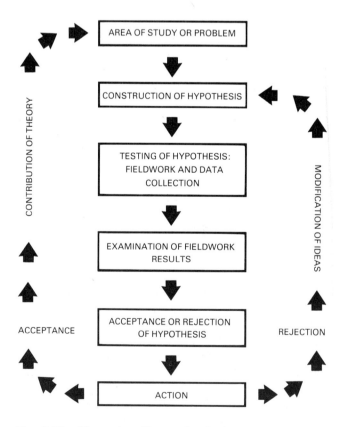

Fig. **2.30** The scientific method of enquiry

Atmospheric Processes   3

The Weather

Table 3.1 shows the weather readings from a school weather station for one week in January and one week in July. The station is about 80 m above sea level and 27 km to the north-west of London.

> 1 From Table 3.1 draw a graph to show the daily maximum and minimum temperatures for each week. Fig. 3.1 shows you how to start.
>
> 2 From the table work out the following:
> a) The mean (average) temperature for each week. Plot this value as a straight line on the graph you have just drawn.
> b) For each week find the difference between the mean for the week and i) the highest temperature of the week, ii) the lowest temperature of the week.
> c) The number of days on which rain fell out of the 14 recorded.

A definition of the term **weather** is: *the physical state of the atmosphere at a particular place at a particular time.* The data in Table 3.1, your answers to exercises 1 and 2 and above all your own experience, tell you how variable and unpredictable the weather can be. The pupils who recorded the weather shown in Table 3.1 had no idea that the bright, frosty start to the week in January would turn into the damp, mild conditions at the end of the week. Perhaps they did make an accurate guess that the hot, sunny weather in July would end in a thunderstorm. There is some meteorological truth behind Henry VIII's statement that an English summer consists of three fine days and a thunderstorm!

The term 'weather' must not be confused with the term 'climate'. **Climate** is the general state and behaviour of the atmosphere over a considerable length of time: it is usually described for a season or a whole year. For example, we can say that Britain has moist, mild winters and moist, cool summers, whereas the Sahara has conditions that are hot and dry throughout the year. Weather readings from a number of stations over a considerable number of years (at least thirty) are required for defining the climate of a given region. Climate is a generalisation of day to day conditions, worked out by taking the mean values from weather readings.

Day	Temperature				Rain	Wind	Pressure	Weather
JAN	I	II	III	IV	V	VI	VII	VIII
Sun	2	-5	7	-1.5	0	NE 1	1020	-1 mist, frost
Mon	1	-8	9	-3.5	0	N 1	1018	-1 clear, frost
Tues	4	-1	5	1.5	0	SE 2	1012	2 cloud 8 oktas
Wed	6	4	2	5	5	SE 6	1000	3 drizzle 8 oktas
Thur	7	6	1	6.5	0	SW 5	998	4 cloud 8 oktas
Fri	9	8	1	8.5	1	SW 4	1000	6 cloud 8 oktas
Sat	10	8	2	9	2	SW 4	1001	8 cloud 8 oktas

JULY								
Sun	26	15	11	20.5	0	S 1	1025	17 mist
Mon	26	12	14	19	0	SE 1	1024	16 mist
Tues	25	10	15	17.5	0	SE 2	1018	17 clear
Wed	22	20	2	21	0	E 3	1015	12 cloud 8 oktas
Thur	20	8	12	14	10	SW 4	1000	10 thunderstorm 7 oktas
Fri	22	10	12	16	2	NW 3	1000	13 cloud 4 oktas
Sat	22	9	13	15.5	1	NW 3	1002	10 cloud 3 oktas

I: maximum
II: minimum
III: diurnal range
IV: daily mean
V: millimetres
VI: direction and force (Beaufort scale)
VII: millibars
VIII: weather recorded at 9 a.m.

Note₁: temperatures in °C
Note₂: one okta is one eighth cloud cover

Table **3.1** Readings from a school weather station

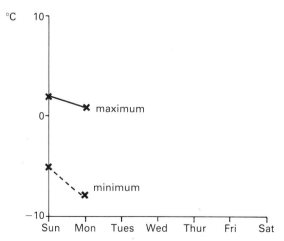

Fig. **3.1** Maximum and minimum temperatures for a week in January

43

We have seen from the readings in Table 3.1 that the temperature during the January week varied a good deal from the week's mean value. The month of January as a whole can vary quite a lot from the average January condition, as shown in Table 3.2.

Year	Mean °C	Climatic mean °C
1976	6.1	4.0
1979	0.85	4.0
1983	8.3	4.0

Table **3.2** Sample January temperature means at London

> **3** From Table 3.2 work out the difference of each January mean from the climatic mean.

In fact, January 1979 was the third coldest this century. Summers can be just as variable. The longest day in 1973 actually recorded a maximum temperature that was *below* the maximum recorded the previous shortest day, December 1972. These examples illustrate that mean figures have to be dealt with very carefully as they do conceal considerable variations.

Measuring and Recording the Weather

Temperature

The maximum temperature over a given period is measured with a **maximum thermometer**, illustrated in Fig. 3.2 (a). The mercury expands in response to a rise in temperature and pushes the indicator up the tube. The indicator is held in position by a spring so that when the mercury contracts on cooling, it remains in position. It must be reset with a magnet after the reading has been made.

The minimum temperature is recorded with a **minimum thermometer**, illustrated in Fig. 3.2(b). This contains alcohol that contracts on cooling, pulling the indicator back with its meniscus. When the alcohol expands as the temperature rises, it flows past the indicator which is held in place by a spring. The indicator must be reset after each reading by using a magnet.

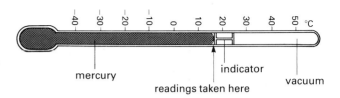

Fig. **3.2 (a)** Maximum thermometer

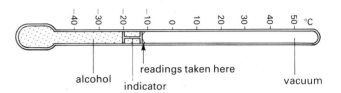

Fig. **3.2 (b)** Minimum thermometer

> **4** After studying Fig. 3.2 explain why it is easy to accidentally *misread* maximum and minimum thermometers by several degrees.

The current air temperature is read from a standard **dry bulb thermometer** which is kept, along with the maximum and minimum thermometers, in a **Stevenson Screen** (Fig. 3.3). The sides of a Stevenson Screen are made of separate slats, or louvres, arranged in two layers. The Screen is placed on a stand which is about 120 cm high. It should be sited on open ground, away from trees and buildings, and with the door facing north.

> **5** Why do you think it is necessary to take temperature readings in the shade?
>
> **6** Explain why a Stevenson Screen is painted white and has louvred sides.

Readings for a particular station can be tabulated, as in Table 3.1, or recorded on a graph, like the one you drew in exercise 1. Graphs can easily be constructed for the whole year, as shown on page 78 where a temperature graph is combined with a rainfall graph. Geographers are particularly concerned with comparing temperatures between different places and analysing the patterns made by temperature changes across regions. Temperatures are mapped by means of **isotherms**, which are lines on a map that join places with the same recorded temperature. Figs. 4.4 and 4.5 show

Atmospheric Processes

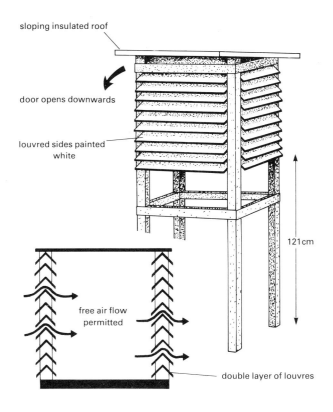

Fig. 3.3 Stevenson Screen

world isotherms at selected values. Two important points need to be made about isotherms in general and the world isotherms in particular. First, they are constructed using *sea level* temperatures, that is, they ignore the effects of altitude. Second, they are highly generalised, especially on the world scale where large areas have no recording stations. The map of Leicester in Fig. 3.18 shows, however, that quite accurate isotherms can be drawn on the micro-scale.

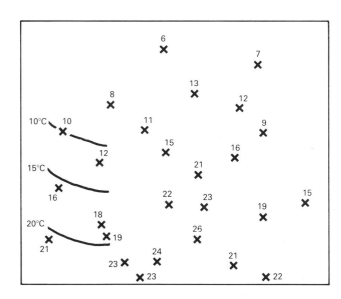

Fig. 3.4 The construction of isotherms at 10°C, 15°C, 20°C and 25°C

Isotherms are drawn at selected intervals from data that is already plotted on a map of the study area. Judgement has to be used when drawing the lines between the known value points of the stations. For example, on Fig. 3.4, the 15°C isotherm goes just on the cold side of the station recording 16°C and mid-way between the next two stations recording 12°C and 18°C. The 20°C isotherm passes between the 16°C and 21°C points, but closer to the warmer of the two. It then skirts round the warmer side of the 19°C point.

> 7 Place a piece of tracing paper over Fig. 3.4 and trace off the points showing temperature and the start of the isotherms shown. Complete the pattern of the isotherms across the rest of the map area. Draw in where you think the 25°C isotherm will go.

Rainfall

Rainfall is collected in a **rain gauge**, illustrated in Fig. 3.5. The water collected over a given period, usually 24 hours, is poured into a measuring cylinder. A reading of, say, 10 mm means that if no rainwater had evaporated, soaked into the ground or run off into streams, the ground would have been flooded to a depth of 10 mm.

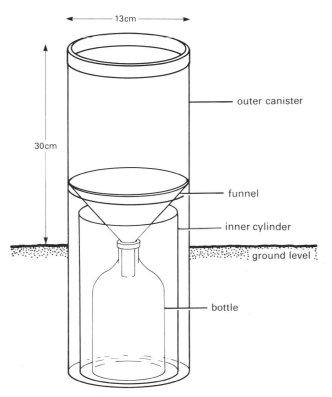

Fig. 3.5 Rain gauge

> **8** As a rain gauge catches free-falling raindrops, explain why care must be taken when siting the instrument in relation to bushes, trees and walls.

Rainfall statistics can be tabulated and graphed, as shown on page 78. They are frequently mapped in a similar way to temperatures: in this case the lines are called **isohyets**.

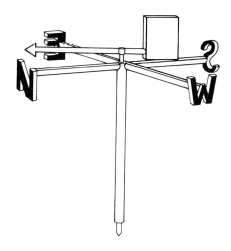

Fig. **3.6** Wind vane indicating a northerly wind

Wind and Pressure

Wind direction is measured by a **wind vane**, often seen on church towers. The arrow head on a vane points *into* the wind, but the direction is described according to the direction *from which* the wind is blowing. So a *northerly* wind comes from the north and is blowing towards the south, as shown on Fig. 3.6.

Wind speed can be measured accurately, but expensively, with an **anemometer**. The most usual type of anemometer consists of three cups fixed to a central vertical shaft which rotates as the cups are pushed round by the wind. This rotation operates a meter which shows the speed of the wind in miles or kilometres per hour, or in knots. A cheaper way to judge wind speed is to consult the **Beaufort Wind Scale** – Table 3.3.

This table shows wind speed in **knots**. One knot is one nautical mile per hour, or 1.86 km/hr. The wind arrows are those used on weather maps issued by the Meteorological Office. We shall be studying these later in the chapter.

Variations in wind speed and direction for one particular place are often recorded in the form of a **wind rose**, like the one shown in Fig. 3.7.

Scale	Description	Effects	Speed (knots)	Arrow indication
0	Calm	Smoke rises vertically	under 1	◎
1	Light air	Direction shown by smoke but not by wind vane	1–3	
2	Light breeze	Leaves rustle, wind vane moved	4–6	
3	Gentle breeze	Leaves and twigs in constant motion	7–10	
4	Moderate breeze	Raises dust and paper; small branches moved	11–16	
5	Fresh breeze	Small trees begin to sway	17–21	
6	Strong breeze	Large branches in motion; whistling heard in telegraph wires; umbrellas used with difficulty	22–27	
7	Near gale	Whole trees in motion	28–33	
8	Gale	Twigs break off trees; difficult to walk	34–40	
9	Strong gale	Slight structural damage, especially to roofs and chimneys	41–47	
10	Storm	Trees uprooted; considerable damage	48–55	▲
11	Violent storm	Widespread damage	56–63	
12	Hurricane	Widespread devastation	above 63	

Table **3.3** The Beaufort Wind Scale

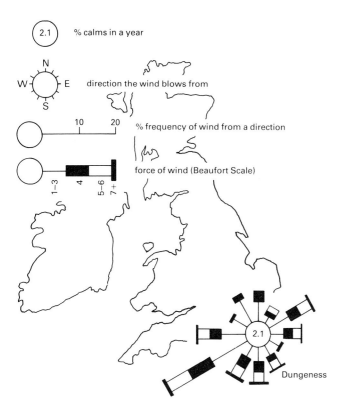

Fig. 3.7 Wind rose for Dungeness

Fig. 3.8 Barograph

> **10** Refer back to Table 3.1. Can you see any relationship between the air pressure recorded and a) wind speed and b) rainfall?

> **9** Study the wind rose in Fig. 3.7. Which is the most common (or **prevailing**) wind direction? Which wind direction is associated with the strongest winds?

A barometer measures **air pressure**. The **aneroid barometer** consists of a collapsible box containing a partial vacuum which responds to changes in air pressure; a needle is attached to this box and shows the pressure on a dial. Some instruments, called **barographs**, have a rotating drum and an ink-loaded needle so that continuous recordings can be made (Fig. 3.8). The traditional barometer (**Fortin barometer**) has a reservoir of mercury. The pressure of the air pushes the mercury up a tube. Average air pressure will support a column of mercury 760 mm high; movement up or down from this level will indicate high or low air pressure. The Meteorological Office measures air pressure in millibars (mb), the average air pressure being 1013.25 mb. (A millibar is equivalent to a force of 100 newtons acting on one square metre.) The lowest recorded pressure in Britain was 925.5 mb on 26 January 1884 at Ochtertyre, Perth. The highest was 1054.7 mb at Aberdeen on 31 January 1902. The usual range is between 990 and 1040 mb.

Air pressure is mapped using lines that join places with the same recorded pressure. These lines are called **isobars**.

Energy in the Atmosphere and Factors Affecting Temperature

The Sun is the atmosphere's only source of energy and is responsible for all weather phenomena. The energy from the Sun is referred to as **insolation**.

How the Atmosphere is Heated

The Earth intercepts only about one two-thousand-millionth of the total energy output of the Sun. The amount of insolation that is received on the Earth's surface from a vertical beam is approximately 1.396 kilowatts per square metre. Rather more insolation than this actually reaches the uppermost layers of the atmosphere, but several things happen to it as it passes through to the ground, as shown in Fig. 3.9.

> **11** From Fig. 3.9 work out the proportion of incoming insolation that is not lost, and which is available for heating the ground.

Figures refer to number of units of energy. **100** units are assumed to enter the atmosphere for the purpose of this diagram.

A: reflection from clouds
B: reflection from ground
C: absorption by gases in the atmosphere
D: scattered by dust particles and water droplets

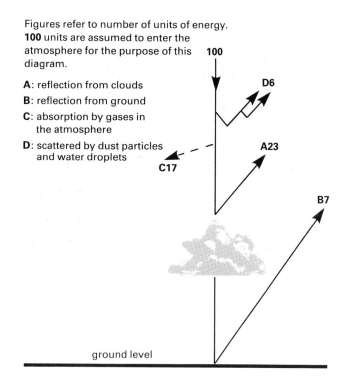

Fig. **3.9** Incoming insolation

The insolation is in the form of short-wave radiation. It does not heat the atmosphere as it passes through it, but the ground is heated by it, and the ground in turn heats the atmosphere. Three processes are involved in the heating of the atmosphere by the ground:
a) long-wave radiation
b) conduction (the transfer of heat by contact)
c) convection and turbulence (warm air moving up and around the atmosphere).

The Effects of the Seasons and Day and Night

Summer is warmer than winter because the Sun is higher in the sky and the length of daylight is longer. The seasonal changes in the altitude of the Sun and the length of day and night are due to the way in which the Earth rotates around the Sun, and around its own axis. Fig. 3.10 shows the orbit of the Earth around the Sun, a journey that takes 365.25 days to complete. An imaginary line drawn between the centre of the Earth and the centre of the Sun for one year would be drawn along the Plane of the Ecliptic.

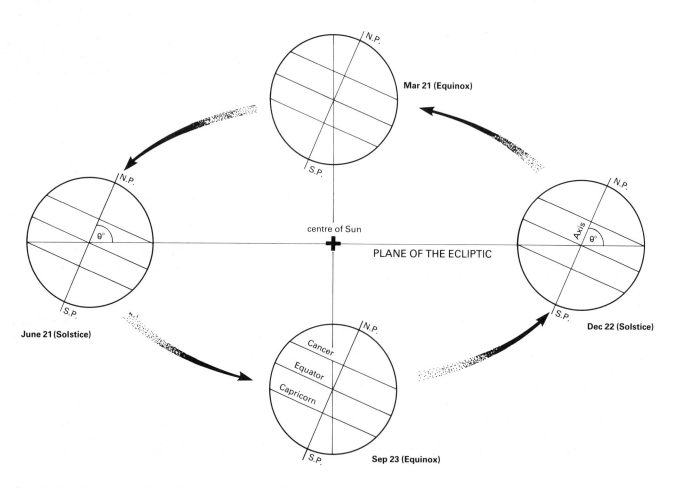

Fig. **3.10** The orbit of the Earth around the Sun

Atmospheric Processes

12 Study Fig. 3.10 very carefully. What is the value of the angle made between the axis of the Earth and the Plane of the Ecliptic? Does this angle of inclination change at all during the year?

13 On 21 March and 23 September the Sun is vertically overhead the Equator at noon local time. Over which lines of latitude is the Sun overhead at noon on a) 21 June and b) 22 December?

14 Make a large, accurate copy of Fig. 3.11 and Fig. 3.12. From each diagram you have drawn find out the following:
a) The angle of the Sun at London.
b) The angle of the Sun over the Arctic and Antarctic Circles and the two Tropics.
c) The proportion of latitude 52°N that is in darkness. Work this out as a fraction of 24 hours to give the length of night in London.
d) The part of the Earth's surface that always has equal lengths of night and day.
e) The parts of the Earth's surface that have six months of daylight and six months of night.

15 Draw a similar diagram to Figs. 3.11 and 3.12 for 21 June.

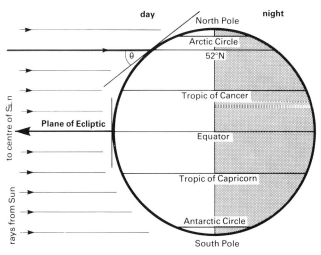

Fig. **3.11** The altitude of the Sun, and day and night, on 21 March and 23 September (the equinoxes)

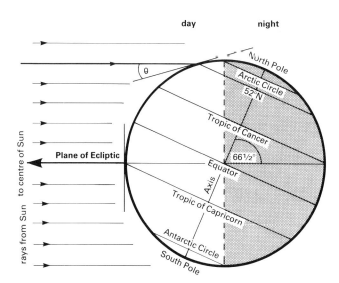

Fig. **3.12** The altitude of the Sun, and day and night, on 22 December (the winter solstice)

As we are on the moving planet, it is the Sun which appears to move throughout the year from Tropic to Tropic and back again.

A more detailed picture of the spring and autumn equinoxes and the winter solstice is given in Figs. 3.11 and 3.12. These diagrams also show day and night. Night is the time during which half of the globe is facing away from the Sun, and is therefore in shadow. During the day, the altitude of the Sun reaches its peak at noon (Greenwich Mean Time) and the angle is measured from the horizontal, or the tangent to the surface of the Earth. On the diagrams the angle of the Sun at noon in London (52°N) is marked by θ.

As the Sun is higher in the sky in summer than in winter, its heating potential is greater, as shown in Fig. 3.13 overleaf.

16 From Fig. 3.13 explain why a higher Sun heats the ground more.

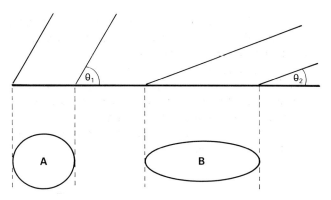

θ₁: angle of Sun's rays in summer
θ₂: angle of Sun's rays in winter
A: area of ground heated in summer
B: area of ground heated in winter

Fig. **3.13** The influence of the angle of the Sun's rays on intensity of heating

The relationship between the angle of the Sun's rays and the potential for heating can be demonstrated by shining a torch at different angles onto a horizontal piece of white paper.

The Effects of Cloud Cover

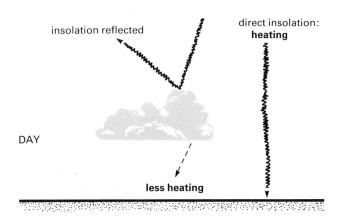

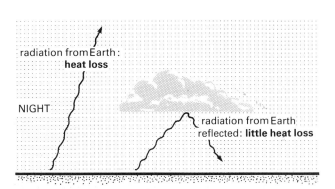

Fig. **3.14** The effects of cloud cover on temperature

17 Refer again to the weather records in Table 3.1. Describe the variation in the **diurnal** (daily) range of temperature for the week in January. Relate this variation to the amount of cloud recorded at 9 a.m.

A cloud cover keeps a night warmer than it would be if the sky were clear. During the day, clouds keep the temperatures lower, particularly in summer. Fig. 3.14 shows why this is so.

The Effects of Altitude and Relief

We have seen that it is the ground that heats the atmosphere and not the Sun: we would therefore expect to find it getting colder the higher up we went, because we would be getting further from the heat source. In addition, the air higher up is thinner and can therefore hold less heat.

18 Test the following hypothesis: *an increase in altitude results in a decrease in temperature*. Table 3.4 gives the altitude and mean annual temperature for a number of stations on or very near the Equator.
Proceed as follows:
a) Plot the altitude and temperature for each station on a **scattergraph**, with temperature along the horizontal axis and altitude along the vertical axis.
b) Draw on the **regression line** (line of best fit), that is, a straight line that has an equal number of points on either side of it.
c) The closer the points are to the line, the better the correlation is between the two variables (altitude and temperature) on the graph. Look at Fig. 3.15 to see whether the correlation on your scattergraph is positive or negative, and strong or weak. Can you reject or accept the hypothesis?
d) Using the regression line, which gives you the *average rate* at which temperature falls with altitude, find out how far the temperature would fall if one climbed to a height of 1 km above sea level.

Atmospheric Processes

	Station	Mean annual temperature (°C)	Altitude (m)
1	Bandoeng	22	730
2	Bogota	14	2661
3	Entebbe	31	1171
4	Fort de Kock	21	1067
5	Georgetown	27	2
6	Iquitos	25	101
7	Kananga	25	627
8	Mombasa	26	15
9	Nairobi	17	1675
10	Ocean Island	27	55
11	Quito	12	2850
12	Singapore	26	3

Table 3.4 Mean annual temperature and altitude of equatorial stations

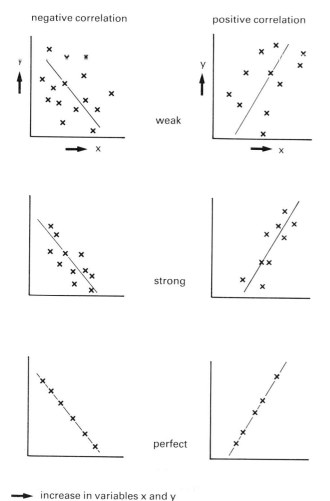

Fig. 3.15 Correlations shown by scattergraphs with regression lines

The fall in temperature with altitude is known as the **environmental lapse rate**. The global mean for this is 6.5°C per kilometre. However, it does vary a great deal, both from time to time and from place to place. In winter in England, for example, the rate can be very much higher than the figure just quoted.

Under certain circumstances, the fall of temperature with altitude can be reversed to form an **inversion** in which the temperature actually rises as you go up. Fig. 3.16 shows the distribution of temperature upwards through the atmosphere.

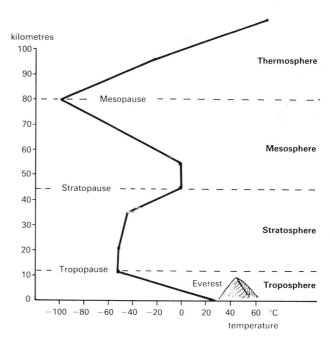

Fig. 3.16 Distribution of mean temperature with altitude

> **19** What happens to the temperature at a height of about 12 km? What is this level called? Between which two layers of the atmosphere does this level occur?

At the **tropopause** there is a very significant inversion. As we shall see later, it acts as a sort of atmospheric 'lid' which prevents hot rising air escaping into the stratosphere.

On a very much more local scale, inversions are frequent occurrences in valley bottoms. Cold air is heavier than warm air. With night-time cooling, cold air will sink down the valley sides, collecting in the bottom. This makes the valley floor colder than higher levels, resulting in a **frost hollow**. One of the most celebrated

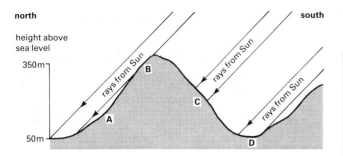

Fig. **3.17** The influence of aspect on temperature

frost hollows occurs in a narrow valley between Rickmansworth and Chorleywood in south-west Hertfordshire. The night climate is one of the most severe in the country. Freezing point has been recorded in July, whilst in January temperatures as low as −19°C have been known.

Aspect, or the direction in which a slope faces, has an important influence on temperature, as illustrated in Fig. 3.17.

> **20** Explain how aspect will have an effect on insolation and temperature at each of the three sites marked A, B and C in Fig. 3.17. Where on this diagram would a frost hollow be found?

The influences that relief has upon temperature do have important implications for farmers and market gardeners.

> **21 a)** What might be the danger for orchards if they were planted in a valley bottom?
> **b)** What differences in crop ripening might be noticeable between sites A and B on Fig. 3.17 and between sites A and C?

Certain weathering and erosion processes are also influenced by aspect.

> **22** Consider how mechanical weathering may be affected by aspect.

The Effects of Settlements

> **23** Fig. 3.18 shows the distribution of temperature across the city of Leicester on one particular night. Describe the pattern shown by this map. How could you account for the pattern you have described?

Fig. **3.18** Leicester's heat island

Leicester is an example of a **heat island** produced by all settlements of every size, although smaller places develop a less pronounced one. Heat islands are particularly noticeable on calm, clear nights, when warm air covers the settlement like a dome.

Water in the Atmosphere

The atmosphere is a mixture of gases. Nitrogen, oxygen, carbon dioxide and water vapour make up 99.997% of it by volume below a height of 90 km. Of these, **water vapour** is responsible for many of our weather features, notably clouds and **precipitation**. The word precipitation covers rain, snow, sleet and hail. All these are the result of water vapour condensing into liquid water or ice. Condensation at ground level produces dew, frost, mist and fog. The amount of water vapour in the atmosphere is extremely variable from place to place and from time to time. It is almost absent above the tropopause, but it can form up to 4% by volume of the air near the surface.

How Water Vapour is Measured

Fig. 3.19 shows a **whirling hygrometer**, consisting of a **wet bulb** and a **dry bulb** thermometer. Water will evaporate from the wet muslin wrapped around the wet bulb (unless the air is saturated with water vapour). Heat energy is extracted from the air around the muslin for this **evaporation** to take place, and so the wet bulb thermometer will record a lower temperature than the dry bulb thermometer. This difference is called the **depression of the wet bulb**. Its value is looked up in hygrometric tables against the dry bulb temperature to find how much water vapour there is in the atmosphere. The amount is expressed in terms of **relative humidity** (the amount of water the air *is* holding as a percentage of the maximum it *could* hold at the same temperature). Warm air can hold more water vapour than cold air. When the temperature falls, the relative humidity may rise to 100%. The air is then said to be **saturated** and **condensation** will occur onto particles of salt and dust.

Fig. **3.19** Whirling hygrometer

The Hydrological Cycle (or Water Cycle)

The amount of water vapour in the atmosphere at any one time is only enough to supply the world with ten days of its normal rainfall. If supplies of water vapour were not constantly added to the atmosphere and recycled, the Earth would very soon dry up. Fig. 3.20 shows how the water is recycled. Eighty-four units of water vapour enter the atmosphere by evaporation from the oceans and seas. Sixteen units enter by evaporation from lakes, rivers and the soil, and by transpiration from plants (water loss from leaves). **Evapotranspiration** is the term used to cover both these contributions (evaporation + transpiration) from the land. The total of 100 units entering the atmosphere is equivalent to the amount of water vapour that would supply the world with 857 mm of rain for a year (i.e. the average annual

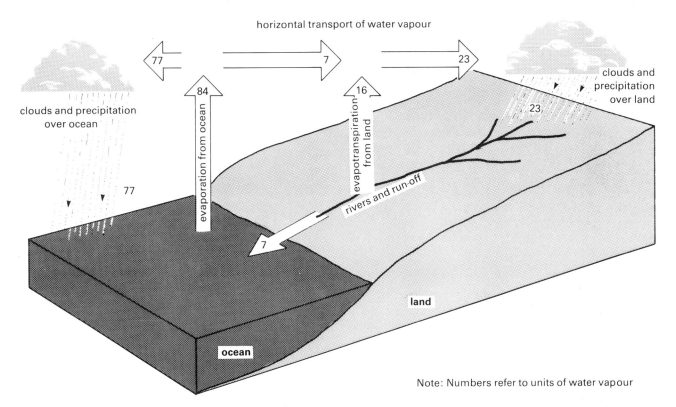

Fig. **3.20** The hydrological (or water) cycle

rainfall). The wind systems of the world are extremely important in the transfer of water vapour around the globe: these will be studied later.

> **24 a)** From Fig. 3.20 work out the proportion of water vapour supplying the rainfall over land that comes from ocean evaporation.
> **b)** What phenomena, particularly in winter, might delay the return of water to the ocean from the land?

Clouds and Precipitation

The cooling of air to its saturation point is the basic cause of clouds, rain and other precipitation. This cooling can be achieved in a variety of ways:
a) by uplift through i) **convection**, ii) rising over high ground, and iii) spiralling upwards in depressions and cyclones;
b) through **radiation** of heat from the ground;
c) by passing over a cold land or sea surface (**advection**).
In each of these cases there must be sufficient water vapour in the air to start with for saturation to take place.

Convection

This is a process in which hot bubbles of air rise upwards. These bubbles are known as **thermals**. Glider pilots and birds use them to gain height; they can also make a jet flight very bumpy. Thermals are formed in the first place because certain types of land surface get hotter than others (arable land gets hotter than woodland for example) and so the air in contact with the hotter surfaces warms up more than the surrounding air. It therefore starts to rise. As the thermals rise they expand, because air pressure exerted on them decreases with height. The rising thermals also cool as they expand, at a constant rate, known as the **dry adiabatic lapse rate**.

> **25** Fig. 3.21 shows the dry adiabatic lapse rate for a rising thermal.
> **a)** Work out the fall in temperature per kilometre.
> **b)** As a rising thermal cools, what will happen to its relative humidity?
> **c)** If the humidity reaches 100% what will happen? (See Fig. 3.22.)

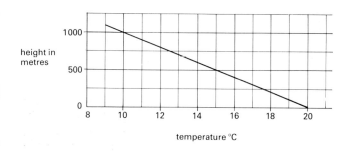

Fig. **3.21** The dry adiabatic lapse rate of a rising thermal

It must be remembered that condensation of droplets will only occur if there are particles like dust and salt in the air onto which the droplets can form. Dust can come from volcanoes and industrial pollution; oceans supply the salt.

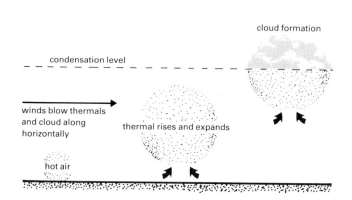

Fig. **3.22** A rising thermal

> **26** Study the cloud photographs in Fig. 3.23. Which type of cloud do you think forms from a rising thermal?

If the cloud grows large enough it will produce a shower. A thermal will only rise if it remains warmer than the atmosphere around it. The atmosphere is then described as **unstable**. If the atmosphere is warmer than the thermal, the thermal cannot rise any further: only small clouds may form, or none at all if the saturation level has not been reached. In such conditions the atmosphere is said to be **stable**. **Cumulonimbus** clouds and **thunderstorms** are the result of vigorous thermals and a very unstable atmosphere. Fig. 3.24 shows the structure of a thunderstorm.

Cloud type A

Cloud type B

Cloud type C

Fig. 3.23 Cloud formations

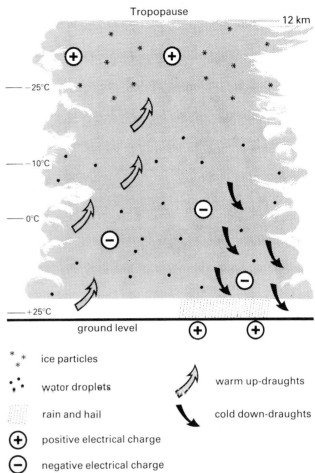

Fig. **3.24** The structure of a thunderstorm

27 a) Why does the upper part of the cloud consist of ice particles? This occurs even in the Tropics.
b) Why does the cloud not rise higher than the tropopause? (Re-read page 51.)

The strong updraughts in the cloud, produced by the rising thermals, carry the water droplets into the top of the cloud, where they freeze; after falling back to the lower part again where they accumulate water droplets, they are carried up again and the new water layer is frozen. This up and down motion eventually creates a lump of ice too heavy to be buoyed up any more and so it falls to the ground as a **hailstone**. Many hailstones melt on the way down to give heavy rain. The usual diameter for a hailstone is between 5 and 50 mm. The largest ones occur in the Tropics and over continents in summer where intense heating causes considerable instability and large cumulonimbus clouds.

The ice particles in the top of the cloud have a positive charge, while the droplets at the bottom have a negative charge. The discharge of electricity between the two creates lightning. Thunder is the sound of rapidly expanding air due to the intense heating from the lightning flash.

Showers and thunderstorms are examples of **convectional rainfall**. Individual showers and storms rarely last more than an hour, but a good deal of rain can fall from them. At Ilkley, 94 mm fell in 1¼ hours on 12 June 1900 (about 10% of the annual total). In Holt, Montana, 305 mm was recorded in 42 minutes on 22 June 1947. The areal extent of a storm is only a few kilometres across, but they can occur in groups which increases both the size of the area affected and the total amount of time they are active.

Relief Rainfall (or Orographic Rainfall)

As high ground, in the path of a stream of air, causes that air to rise and therefore cool down, we might expect to find that rainfall totals are greater on high ground than on low ground.

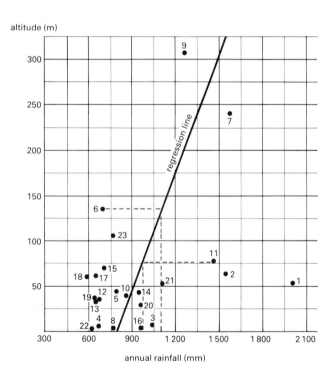

Fig. **3.25** Correlation between altitude and rainfall

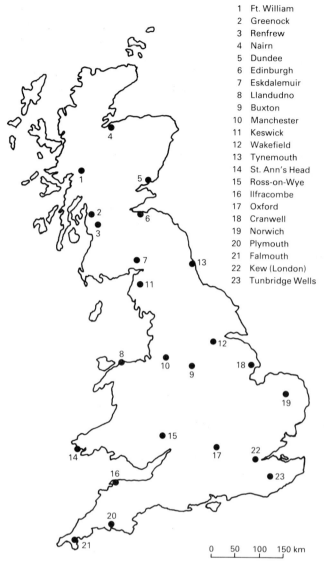

1 Ft. William
2 Greenock
3 Renfrew
4 Nairn
5 Dundee
6 Edinburgh
7 Eskdalemuir
8 Llandudno
9 Buxton
10 Manchester
11 Keswick
12 Wakefield
13 Tynemouth
14 St. Ann's Head
15 Ross-on-Wye
16 Ilfracombe
17 Oxford
18 Cranwell
19 Norwich
20 Plymouth
21 Falmouth
22 Kew (London)
23 Tunbridge Wells

Fig. **3.26** Distribution of rainfall stations plotted on Fig. 3.25

There is clearly a positive correlation between altitude and rainfall shown on the graph, but there are some points on the graph that are rather a long way from the regression line. This means that there is another factor (or group of factors), apart from altitude, affecting rainfall.

28 Test the hypothesis that *as altitude increases so rainfall will also increase*. Fig. 3.25 is a scattergraph of altitude plotted against mean annual rainfall for a number of stations in Britain. The map in Fig. 3.26 shows their geographical distribution. What conclusion do you come to about the validity of our hypothesis? (Refer to Fig. 3.15 for the interpretation of scattergraphs.)

29 Locate station 6 on the graph. Its rainfall is 699 mm. By following the dotted line from point 6 on the graph along to the regression line and then vertically down to the *x*-axis we can read off the amount of rainfall the station *ought* to have for its height. What is this value?

Station 6 is a **negative residual** because it has a lower rainfall than predicted by its altitude. All the points to the left of the line are negative residuals. In a similar way we can demonstrate that all the points to the right of the line are **positive residuals**, that is, they have a rainfall that is higher than we would predict from their altitude.

> **30** Follow the dashed line from point 11 on the graph to find out how much rain we would *expect* to find there.
>
> **31** Make a copy of the map in Fig. 3.26. Circle all the stations that are negative residuals with a single ring and the stations that are positive residuals with a double ring. Describe the distributions of these positive and negative residuals.

The distribution pattern you have observed can be explained by the fact that the most common wind direction (or prevailing wind) over Britain is between north-west and south-west, off the Atlantic Ocean. These are our main rain-bearing winds, therefore the stations exposed to them on the western side will receive more rainfall than those on the eastern side. This fact weakens the simple correlation between altitude and rainfall.

The air that rises over high ground cools by expanding, like a thermal. However, the air rises not in bubbles but in layers, creating layered or stratiform clouds, provided there is enough water vapour available. The names given to such clouds are **stratus** and **nimbostratus**.

The cloud base may come right down to ground level, giving hill fog. Rain and drizzle can last for several hours or even days, with snow in winter if it is cold enough.

However, the *intensity* of precipitation is generally much less than from convectional processes.

On the leeward side of the high ground (that is, the side that faces away from the oncoming wind) it is drier because the air begins to descend, contract, and become warmer: the cloud will start to evaporate as the relative humidity goes down. The leeward side is said to be in a **rain shadow**.

> **32** From the photographs of cloud types in Fig. 3.23, identify that which is produced by air rising over high ground.
>
> **33** Make a copy of Fig. 3.27 and provide suitable labels to describe the processes that are operating at each of the lettered points A–D.

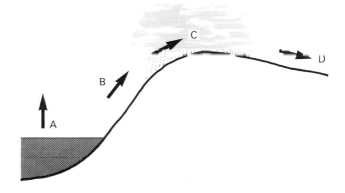

Fig. 3.27 The formation of relief rainfall and a rain shadow

Cyclonic and Frontal Rainfall

Cyclones and **depressions** are areas of low atmospheric pressure in which the air is spiralling upwards. This

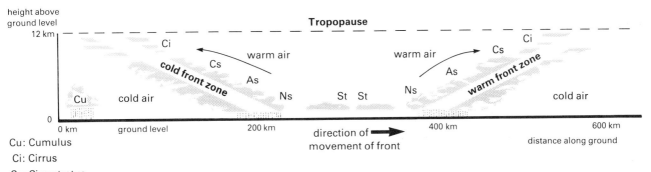

Cu: Cumulus
Ci: Cirrus
Cs: Cirrostratus
As: Altostratus
Ns: Nimbostratus
St: Stratus

Note: Only one kind of warm and cold front is shown here. Other types do frequently occur over Britain. The angle of the fronts with the ground has been exaggerated.

Fig. 3.28 The formation of frontal rainfall

upward movement means that they are associated with rain, often persistent and heavy. Cyclones are found in the Tropics and are the result of intense instability in hot, moist air. Depressions are found in higher, more temperate latitudes where cold air from the Poles meets warm air from the Tropics. Warm and cold air do not mix: the warm air rises over the cold along **fronts**, as shown in Fig. 3.28. The formation of depressions and fronts will be studied in a later section; it is necessary at the moment to understand that warm air, rising over the cold air along the warm and the cold fronts, causes the same processes to occur as when air rises up the windward side of high ground.

> **34** Study Fig. 3.28.
> a) What evidence is there to indicate that frontal rainfall is similar to orographic rainfall?
> b) What evidence is there to indicate that *convectional* rainfall occurs behind the cold front?

Convection occurs behind the cold front because the polar air is cold enough to allow thermals to rise upwards.

Other Forms of Precipitation
Rainfall totals for weather stations also include snow, sleet, hail and drizzle. **Snow** is made up of hexagonal ice crystals that form in a cloud when the temperature falls below $-10°C$. (Between $0°$ and $-10°C$ droplets remain liquid, and are known as supercooled droplets.) If the temperature of the air between the cloud and the ground remains below $3°C$ the flakes will survive as they fall, but the ground temperature must be at or below freezing point for the snow to accumulate. High ground has a much greater incidence of snow than low ground.

> **35** Give two reasons why this should be so.

It has been estimated in Britain that for every 30 metre rise in altitude there is one more day per year with snow falling.
  Sleet is a mixture of rain and snow which occurs when temperatures at ground level are between $2°C$ and $4°C$.

Ground level Condensation
A clear night sky permits the loss of radiation from the ground, which causes the ground surface to become colder. The air in contact with the ground will also become colder, often to the point where saturation and condensation will occur: this condensation produces **dew**. If the dew-point is below freezing, or the temperature falls below freezing after condensation, then ice crystals will form to give white or **hoar frost**. **Air frost** occurs when the air temperature above the ground (measured in a Stevenson Screen) falls to $0°C$ or below. **Ground frost** occurs when the temperature of the ground surface falls to $0°C$ or below but the air remains above freezing point.
  Fig. 3.29 shows how mist and fog may form. The difference between mist and fog is only one of visibility: in mist you can see further than 1000 metres.

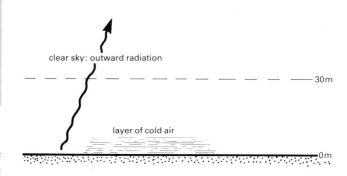

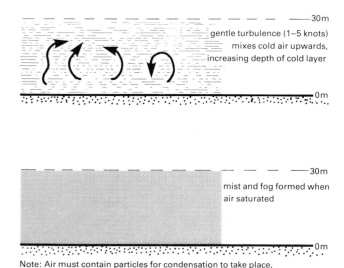

Note: Air must contain particles for condensation to take place, e.g. onto dust and salt.

Fig. **3.29** The formation of mist and fog

> **36** Study Fig. 3.29 and describe in your own words how the fog is formed. (The diagram shows **radiation fog**.)

Atmospheric Processes

In industrial cities, the air needs to reach only 80% relative humidity before condensation begins, because of the pollutants like sulphur dioxide that are in the air. A mixture of fog and pollutants is called **smog**. One of the worst smogs in Britain occurred over London between 5 and 9 December 1952 when up to 4000 people died from its effects. Since the Clean Air Act of 1956 there has been a significant decrease in the number of smogs and fogs in this country.

Freezing fog exists when the air temperature is below 0°C. When the fog droplets touch a cold object like a branch, or a fence, they freeze: a build-up of ice particles in this way is called **rime**.

Advection fog is caused by the movement of warm, moist air over a cold surface (advection means heat transfer by horizontal movement). The Newfoundland and Californian coasts are notorious for this where warm air moves over a cold offshore current. In California the problem is made worse by the mixing in of car exhaust fumes.

Atmospheric Pressure and Winds

If you press down the valve of a car or bicycle tyre, the air will rush out: it moves from the high pressure in the tyre to the relatively low pressure outside. In the atmosphere, wind is caused by air moving from areas of high pressure to areas of low pressure. The difference in pressure is called the **barometric gradient**. Just as contour lines on a relief map get closer together as a slope increases in gradient, so isobars on a weather map get closer together as the barometric gradient increases. An increase in barometric gradient means an increase in wind speed.

Anticyclones

These are areas of high pressure. Fig. 3.30 shows the theoretical structure of an anticyclone in cross-section and how it might appear on a map.

> 37 From Fig. 3.30 give a reason why pressure is high. What is the ground-level movement of the air indicated on the diagram?

However, Fig. 3.30 does not take into account the effect that the rotation of the Earth on its axis has on the air flow. **Ferrel's Law** states that the rotation of the Earth causes air to be deflected to the right in the northern hemisphere and to the left in the southern hemisphere. (Right and left refer to a person standing with their back to the wind.) The deflection of winds like this has a very important effect on our anticyclone, as shown in Fig. 3.31.

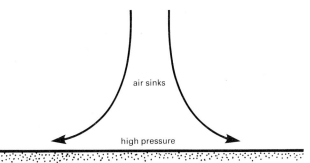
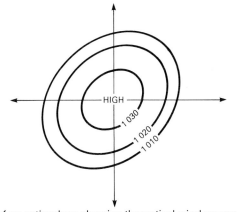

Fig. **3.30** Theoretical air movement in an anticyclone

> 38 Study Fig. 3.31 carefully and answer the following:
> a) Describe the wind motion around the anticyclone.
> b) Is this anticyclone in the northern or southern hemisphere?
> c) What do you notice about the spacing of the isobars as the distance from the centre of the anticyclone increases?
> d) What will happen to the wind speed as distance from the centre increases?

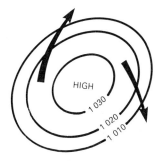

Fig. **3.31** A model anticyclone

As the air sinks in an anticyclone, it will become compressed and warmer.

> **39** As the air sinks, what will therefore happen to the relative humidity? What effect will this have on cloud formation and the likelihood of rain?

We have now built up a picture of the model anticyclone. It is an area of high pressure, around which winds move in a clockwise direction (northern hemisphere), accelerating as they get further out. The weather will be dry and clear. We shall now test this model against a real example. Fig. 3.32 is a weather map (or **synoptic chart**) for 15 February 1981, at 0600 hours (6 a.m.).

> **40** Study Fig. 3.32 and answer the following:
> a) What evidence of an anticyclone is there?
> b) Where is it centred?
> c) Does the wind speed and direction recorded for London, Edinburgh and Stornoway support our model?
> d) Describe the distribution of cloud shown on the map: does this pattern support our model?
> e) What is the particular weather condition reported at London?
> f) Why should London have this particular weather?

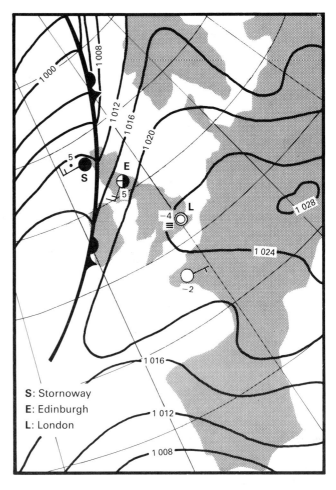

Fig. **3.32** Synoptic chart for 0600 hrs, 15 February 1981

Anticyclones are often associated with radiation fog, although it must be remembered that fog-forming conditions are found sometimes in low pressure areas as well, especially after rain or with thawing snow. The clear skies associated with anticyclones mean cold weather and frost in winter, but warm conditions in summer. In Britain our extreme frosts and heatwaves are normally brought by high pressure areas. As we shall see in the next chapter, the arid condition of deserts is caused by persistent high pressure. Elongated areas of high pressure extending from an anticyclone are called **ridges**: one can be seen on Fig. 3.32 to the west of Brittany.

Cyclones and Depressions

These are areas of low pressure in which air spirals upwards and inwards in an anticlockwise direction in the northern hemisphere, and clockwise in the southern hemisphere. The winds are much stronger than in anticyclones. The upward motion of the air causes outbreaks of rain, showers and thunderstorms. Along the fronts there is more continuous precipitation.

Mid-latitude Depressions and Fronts

Warm and cold air do not mix: they meet along boundary zones called fronts. Between approximately 40° and 65° north and south of the Equator cold air from the Poles meets warm, moist air from the Tropics along the **Polar Front**. The Front is not static, and Fig. 3.34 illustrates what happens along it: this is the formation of a depression with its associated fronts.

> **41** Study Fig. 3.34.
> a) Describe what happens to i) the movement of the warm and cold air and ii) the air pressure and isobars during the development of the depression.
> b) Write a definition of warm front, cold front and occluded front.

Atmospheric Processes

Wind speed and direction

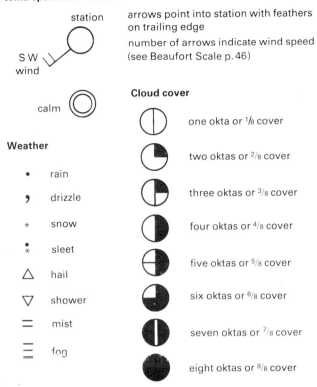

Weather

- • rain
- ❜ drizzle
- * snow
- ⁎ sleet
- △ hail
- ▽ shower
- = mist
- ≡ fog

Cloud cover

one okta or 1/8 cover
two oktas or 2/8 cover
three oktas or 3/8 cover
four oktas or 4/8 cover
five oktas or 5/8 cover
six oktas or 6/8 cover
seven oktas or 7/8 cover
eight oktas or 8/8 cover

Summary for a weather station

north-east wind, force 4 (11–16 knots); 7/8 cloud cover; rain shower; temperature of 9°C

Fig. **3.33** Key to weather symbols on a synoptic chart

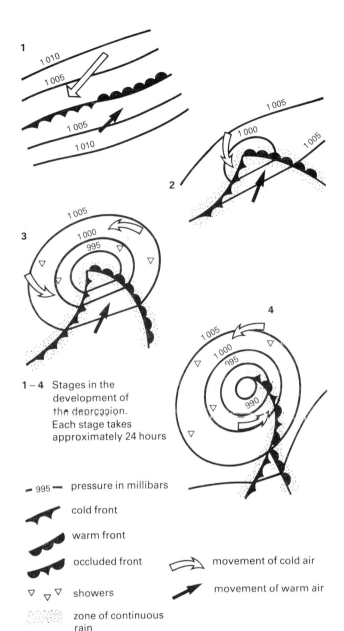

1–4 Stages in the development of the depression. Each stage takes approximately 24 hours

— 995 — pressure in millibars
cold front
warm front
occluded front
▽ ▽ ▽ showers
zone of continuous rain
⇨ movement of cold air
➤ movement of warm air

The area between the warm and cold fronts is known as the **warm sector**, but this is lifted off the ground when the occluded front is formed. The occlusion is formed because the cold front moves faster than the warm front, eventually catching it up.

42 Which one of the photographs in Fig. 3.23 indicates the approach of a front? (Look at Fig. 3.28 to help you answer this.)

Fig. 3.34 is a model of the evolution of a depression. How well does the model depression fit the real-life one? Fig. 3.35 shows the synoptic situation for 6 June 1977, at 1200 hours (noon).

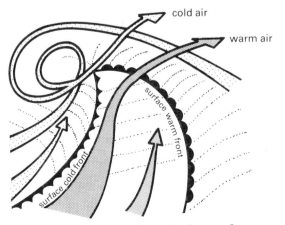

3-dimensional model of depression in stage **3**

Fig. **3.34** The development of a depression

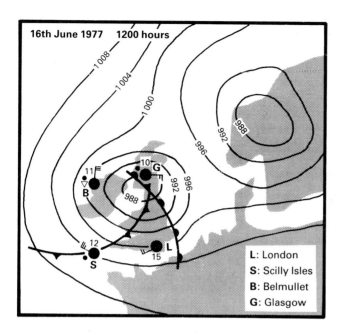

Fig. 3.35 Synoptic chart for 1200 hrs, 6 June 1977

Tropical Cyclones

These originate over the oceans between 5° and 15° north and south of the Equator. They are called **hurricanes** in the Atlantic and **typhoons** in the Pacific. A cyclone is an intense upward spiral, or 'vortex' of warm air producing cumulonimbus clouds and very heavy rain (over 20 mm per hour). In the centre is the 'eye' in which air descends and calm, dry conditions prevail. Around the eye wind speeds can reach 360 km/hr. With these conditions it is not surprising that they wreak havoc in areas affected by them. One of the more notable ones was the Cambay cyclone which crossed the west coast of India near Port Cambay (to the north of Bombay) on 4 June 1976. Seventy people lost their lives and 4500 head of cattle were drowned. Fig. 3.36 shows the wind speeds recorded on the Hakon Magnus and Shenon Doah oil rigs, both 150 km off the coast of India.

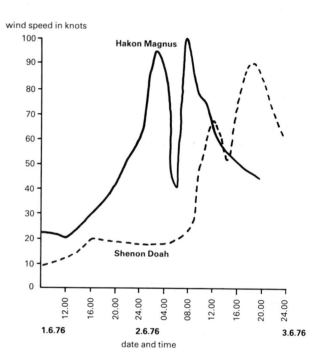

Fig. 3.36 Wind speeds recorded on oil rigs during the passage of the Cambay cyclone, June 1976

> **43** Answer the following questions after studying Fig. 3.35:
> **a)** What is the lowest air pressure indicated *over Britain*? Where is the depression centred?
> **b)** What is the strength and direction of the wind at each of the stations labelled on the map?
> **c)** Why is it raining in Glasgow and in the Scilly Isles?
> **d)** Give the main reason why it is colder in Belmullet than in London.
> **e)** It is showery over Ireland: why?
> **f)** If the depression moves east across the North Sea during the next twelve hours, describe the changing weather conditions over London during this time in terms of i) wind direction, ii) wind speed, iii) cloud and precipitation and iv) temperature.

Not all depressions fit the model by any means. One such example can be seen on Fig. 3.35 to the north-east of Scotland.

> **44** Why do you think that this depression does not fit the model?

> **45 a)** What evidence on Fig. 3.36 is there to suggest that the oil rigs were right in the path of the cyclone?
> **b)** What was the maximum wind speed recorded?
> **c)** If the start of the cyclone is marked by a wind speed of 50 knots, how long did it take for the storm to pass over the Hakon Magnus rig?
> **d)** If the two rigs are 75 km apart, at what speed was the cyclone travelling?

Atmospheric Processes

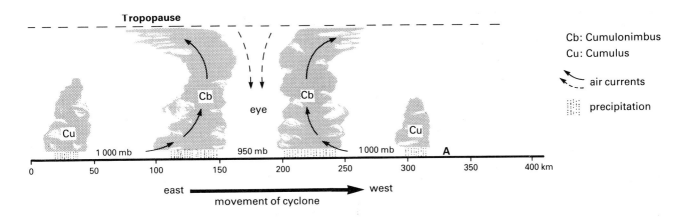

Fig. 3.37 Cross-section through a tropical cyclone

The energy of the cyclones comes from the heat that is released when condensation occurs. This relies on plentiful supplies of warm, moist air from the ocean: as they move over land, this supply is cut off and the storm gradually dies out.

> **46** Fig. 3.37 is a cross-section through a model tropical cyclone. Describe the changes in the weather conditions that an observer at point A will see as the storm passes over.

Local Winds

Included in this category are winds that occur over very restricted areas and for only short periods of time. The relative positions of depressions and anticyclones determine wind direction, as shown in Fig. 3.38. The weather conditions brought by such winds depend upon the season, the direction from which they come and the particular features of the local environment. Fig. 3.39 shows local winds in the Mediterranean area which may blow in certain seasons of the year.

> **47** Each of the winds shown in Fig. 3.39 is associated with particular conditions: a) cold with frost, b) warm with snow-melt, c) dusty and hot. Link each of the winds shown on the map with one of these descriptions. Justify your choices.

The **Chinook** is a wind that blows from the west down the eastern side of the Rocky Mountains onto the Prairies of North America. In winter it can raise the temperature by 20°C in 15 minutes, giving frost-free spells of weather. The name is actually Red Indian for 'snow-eater'.

Fig. 3.38 Distribution of pressure and wind directions in the northern hemisphere

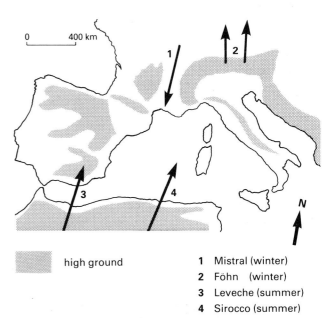

Fig. 3.39 Local winds in the Mediterranean region

1 Mistral (winter)
2 Föhn (winter)
3 Leveche (summer)
4 Sirocco (summer)

> **48** Explain why the descent of the Chinook causes warmer weather. Which wind shown in Fig. 3.39 most resembles the Chinook?

On a smaller scale are **land and sea breezes** which blow in coastal areas. During the day, the sea does not heat up as much as the land because a) a proportion of the Sun's energy is taken down to greater depths in the sea by its transparency and by the action of waves and currents, and b) it takes more energy to heat up a given volume of water than to heat up the same volume of land. Thermals rising over the land create slightly lower pressure than over the sea. Thus an onshore, or sea breeze, develops. At night the land becomes cooler than the sea, the air sinks and moves seawards: a land breeze.

the air pressure can drop by over 150 mb in a few minutes, causing objects to be literally sucked upwards. In 1931 in Minnesota, USA, a tornado lifted an 82 tonne railway carriage with its 117 passengers 25 metres into the air. The violent, twisting winds (up to 500 km/hr) do considerable damage, but over very small areas. The Great Plains of North America in summer probably have the greatest frequency of tornadoes, but they do occur all over the world, including Britain. Here they are usually associated with deep depressions and with warm, unstable conditions in summer: it is estimated that as many as 50 tornadoes can occur every year in Britain. Being small, short-lived phenomena, recording and measuring them is quite difficult.

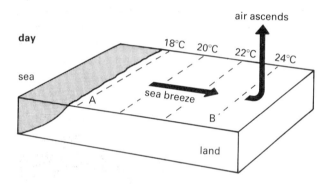

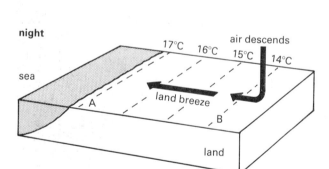

Fig. 3.40 Land and sea breezes

> **49** Fig. 3.40 shows the operation of land and sea breezes. Compare the diurnal range of temperature at A with the diurnal range at B.

Tornadoes, or whirlwinds, are very localised phenomena. They are intense upward spirals of air, only a few hundred metres in diameter. As they pass over,

Workshop

1 Keep a weather log, like that shown in Table 3.1, for your school or home. It should be kept up for at least a season or for a whole year. It may not be possible for you to invest in a Stevenson Screen or maximum and minimum thermometers, but quite valuable observations can be made using just an ordinary thermometer sited in a shaded spot. Readings should be taken three or four times in 24 hours.

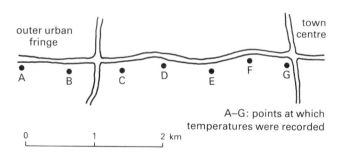

Table of results

Point	Time				
	18.15	18.30	18.45	19.00	19.15
A	1.0	0.5	−1.0	−1.5	−1.5
B	1.0	0.5	0	0	−0.5
C	1.5	1.0	0.5	0.5	0

Date 15/2/83
Weather CLEAR & CALM

Fig. 3.41 Recording an urban heat island along a road

Atmospheric Processes

2 Collect the daily weather report and map for a year. These are obtainable from the London Weather Centre, but a reliable map appears in *The Times, The Guardian* and *The Daily Telegraph* newspapers, with London and Manchester readings. Compare your own records with those published: how do they differ? If they do differ quite a lot, can you explain why? Try to relate the weather you record with the changing pattern of anticyclones and depressions over and around the British Isles.

3 With the help of other members of your class, measure the effect that your local urban area may have upon the temperature distribution at various times of the day and under different weather conditions. Early morning and mid-evening may give a stronger heat island effect (page 52) than at mid-day. Is the effect of the urban area greater or less when it is a) clear or b) cloudy? The most convenient way to measure these effects is to put five or six people, equally spaced, along a 4 km stretch of road leading into the town. Each person should have, ideally, a whirling hygrometer, which is illustrated in Fig. 3.19. This instrument will allow you to compare relative humidities along the road. If you cannot obtain a hygrometer, you will have to restrict your observations to temperature, using standard thermometers. At a given number of specific times, each person reads and records the temperature. A sample set of results is shown in Fig. 3.41.

4 A similar study can be carried out across a local valley: again, interesting differences may be recorded at different times of day and under different weather conditions. How can you explain any differences you record?

5 Analyse the effect your school building may have on a) temperature distribution in the grounds (the school's own heat island) and b) on wind direction. Local winds are very often different from those forecasted nationally. Observations of this can be made with the help of pieces of paper dropped on the ground or with talcum powder puffed from a container.

6 Study Fig. 3.42 which is a satellite photograph taken above Western Europe. Draw a map of the cloud patterns shown. What can be learnt from this pattern about the pressure, winds, and likely weather conditions on the ground? As cloud reflects the Sun's rays, cloudy areas appear white on the photograph, and clear areas appear dark grey.

Fig. **3.42** Satellite photograph over Western Europe (the continental outline is marked to aid location)

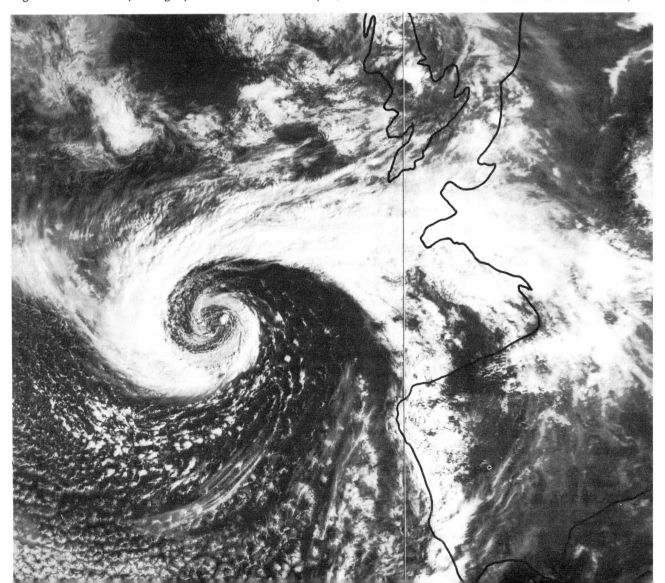

World Climates, Vegetation and Soils 4

Every day a place will experience weather, but the climate can only be understood over a very much longer time span: climatic statistics are usually worked out over a period of at least 30 years. The world may be divided into a number of **climatic regions**, each with its own pattern of rainfall, temperature, pressure and winds. The natural vegetation and soils of each of these regions are closely related to the climate to form a **natural region**.

People will respond to the natural environment in such a region. Just how they respond will depend not only on the natural conditions themselves, but also upon their level of technology and the way in which they organise their society. In Western Europe, the natural vegetation is forest, but most of this has been cleared to make way for commercial agriculture, cities and industry. Human activities have affected the climate, and many of the soils have been altered by the addition of fertilisers. By contrast, the Amazon Basin still has thousands of square kilometres of natural forest, occupied by scattered tribes. These tribes do not support themselves with agriculture, but by collecting berries and fruits and hunting small animals. Fig. 4.1 shows how the various elements relate to each other in a natural region.

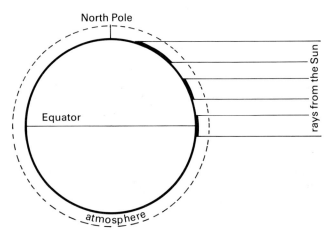

Note: The thickness of the atmosphere has been exaggerated.

Fig. **4.2** The effect of the Earth's curvature on the amount of insolation received

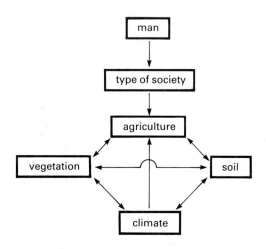

Fig. **4.1** Relationships between the elements of a natural region

1 Describe the relationship between the curvature of the Earth's surface and the amount of insolation received.

2 Fig. 4.2 shows another reason why high latitudes are colder than low latitudes: what is this? (Hint: the atmosphere.)

3 Make a copy of Fig. 4.2, and on it draw the pattern of isotherms (p. 44) that you think would occur as a result of the insolation received. No values need be given.

The Earth's surface loses heat through long-wave radiation. The difference between the incoming (short-wave) radiation from the Sun and the outward radiation from the Earth is called the **heat budget**.

World Climate

Temperature Distribution

We have already studied how the angle of the Sun's rays influences the temperature of a place (pages 48–50). Fig. 4.2 shows how the curved surface of the Earth affects the intensity of insolation.

4 Study Fig. 4.3 carefully. What areas of the Earth have a heat loss over the year as a whole? What areas have a heat gain? Give reasons for your answers.

World Climates, Vegetation and Soils

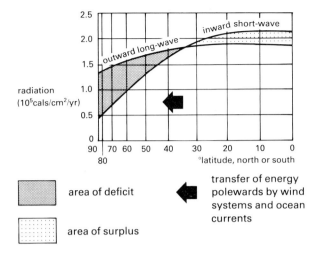

Note: The 'y' axis shows the number of calories that are received on each square centimetre of the Earth's surface each year. A calorie is the amount of heat required to raise 1 g of water by 1°C.

Fig. **4.3** The balance between inward radiation from the Sun and outward radiation from the Earth

Theoretically, the high latitudes ought to get colder and colder each year, while the low latitudes ought to get warmer and warmer. In practice, as we know, this does not happen because a) warm air often blows from the Tropics towards the Poles and b) ocean currents distribute heat energy polewards.

Our conclusions from Figs. 4.2 and 4.3 suggest that the Equator is the hottest place on Earth: in fact this is not so.

5 Figs. 4.4 and 4.5 show the path of the **Thermal Equator**. This is the line that links up the hottest places at a particular time. Describe the path that this line takes in January and July.

6 Study Fig. 4.6 and then in your own words give reasons for the position of the Thermal Equator in January and July.

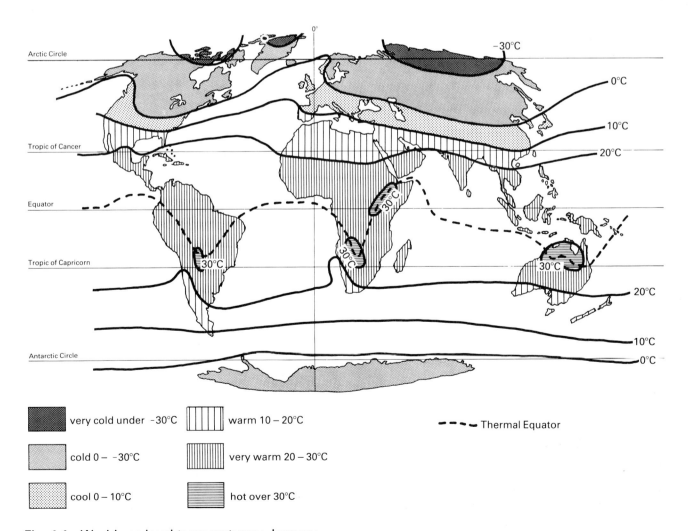

Fig. **4.4** World sea level temperatures: January

Effects of Land and Sea upon Temperature

7 Study the world isotherm maps in Figs. 4.4 and 4.5.
a) Describe the course of the 0°C isotherm in January and the course of the 20°C isotherm in July in each hemisphere.
b) Can you give a reason why the course of the two isotherms you have just described differs from the theoretical pattern you drew in exercise 3?

The land heats up and cools down much more quickly than the sea. Read again p. 64 to see why this is so. Distance from the sea affects both the diurnal range and the annual range of temperature.

Station	Jan. °C	Jul. °C
Shannon (S.W. Eire)	5	16
Warsaw	− 3	19
Aralsk (USSR)	−12	25

Table **4.1** Maximum and minimum temperature means for selected stations

8 Consider the data in Table 4.1. Work out the annual range of temperature for each station. Using your atlas, find out how far each station is from the nearest ocean.

9 Study the map of annual temperature ranges in Fig. 4.7. Give suitable geographical (regional) names to the areas that have a) an annual range of more than 30°C, and b) an annual range of less than 3°C.

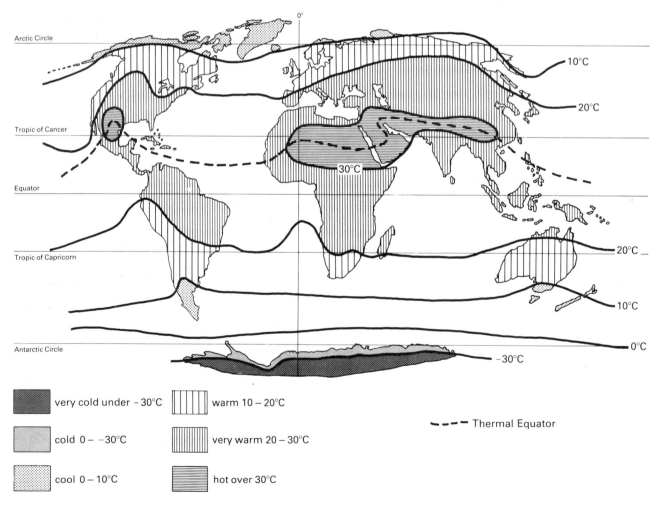

Fig. **4.5** World sea level temperatures: July

World Climates, Vegetation and Soils

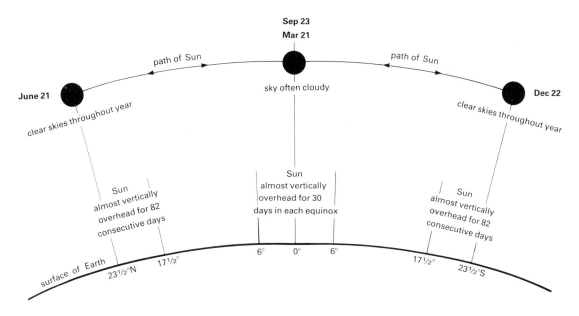

Fig. 4.6 The apparent movement of the overhead Sun during the year

Fig. 4.8 shows the distribution of the major **ocean currents**. These currents are formed by the prevailing winds that pass over the oceans: the prevailing wind direction being the most frequent direction. However, unlike winds, the direction of an ocean current is described according to the direction *towards which* it is moving. For example, the Peruvian Current is a *northerly* current because it flows northwards along the west coast of South America. Cold currents distribute cold water towards low latitudes, and warm currents distribute warm water towards high latitudes.

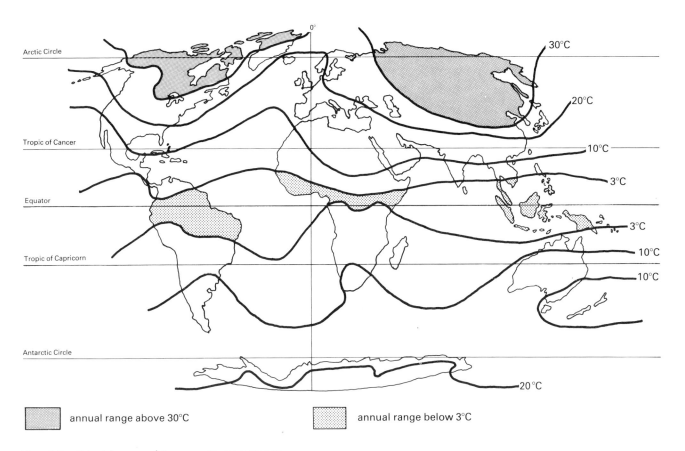

Fig. 4.7 World annual temperature ranges

70 World Climates, Vegetation and Soils

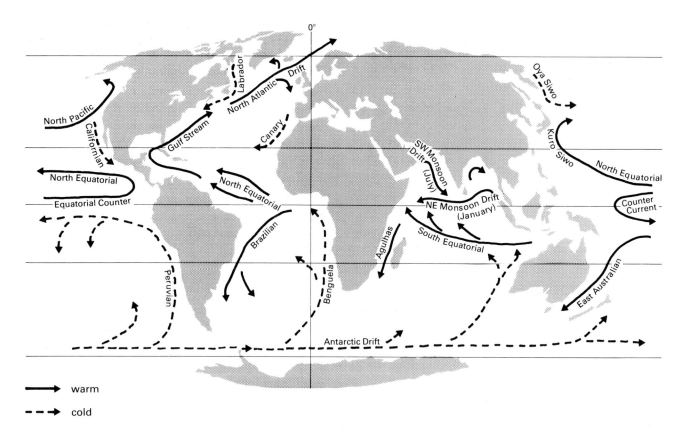

Fig. 4.8 World ocean currents

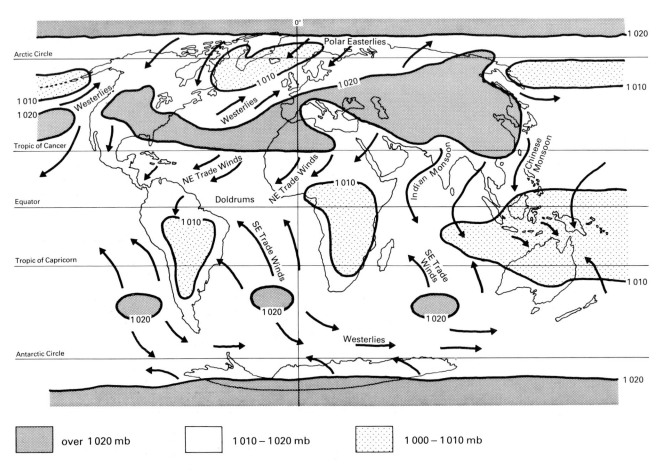

Fig. 4.9 World pressure and winds: January

World Climates, Vegetation and Soils

> **10** Study Fig. 4.8. Using this and the world isotherm maps, describe the effects that the following ocean currents have upon adjacent land areas in January and July: the North Atlantic Drift, the North Pacific Current, the Agulhas Current, the Peruvian Current, the Benguela Current and the Labrador Current. In which season does the influence of each seem to be most marked?

Where winds are onshore, the landward influence of currents can be considerable, but with offshore winds the landward effect is far less. Mountain ranges also limit the influence of currents: notice for example the effect of the coastal mountains in the north-west USA.

Distribution of Winds and Pressure

The distribution of winds and pressure around the globe is shown in Figs. 4.9 and 4.10; Fig. 4.11 shows the paths taken by storm tracks. It is important to remember that the patterns on these maps show the *average* state of affairs. Pressure and winds are dynamic, that is, they are constantly changing.

The most stable elements are the **subtropical high pressure cells** which lie approximately between latitudes 15° and 35° north and south. The areas affected by these cells are known as the **Horse Latitudes**.

> **11** Study Figs. 4.9 and 4.10 very carefully and answer the following questions:
> a) Describe the distribution and extent of the subtropical high pressure cells, noting especially the changes that take place from January to July.
> b) What is the level of pressure between these high pressure cells in the equatorial regions?
> c) Give the direction of the winds that blow on the equatorial sides of the high pressure cells. How can you account for these winds? What are the names given to these winds?

The belt of low pressure around the Equator is a result of convection: warm rising thermals. The regions that experience this are named the **Doldrums**.

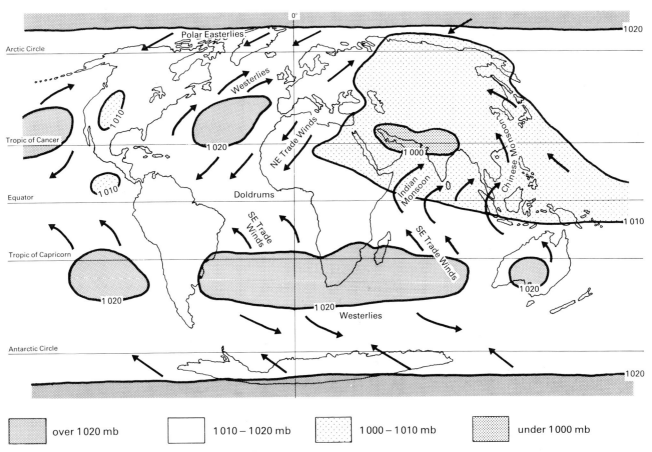

Fig. **4.10** World pressure and winds: July

12 Give the direction of the winds that blow on the poleward sides of the subtropical high pressure cells.

13 What is the level of pressure over the polar regions? Describe the direction of the winds that blow from these regions in each hemisphere.

As we saw in the previous chapter, warm and cold air do not mix: they meet along a front. As the warm Westerlies meet the cold Easterlies from the Poles, the Polar Front is formed.

14 What important atmospheric feature is formed along the Polar Front? (Refer to Fig. 4.11 and pages 60–1).

The large land areas of North America and Asia tend to develop their own pressure patterns in summer and winter.

15 Refer to Figs. 4.9 and 4.10. What pressure levels are found in Asia and North America in a) January and b) July?

The low pressure in July (summer) forms because the land surface becomes hot, and rising thermals develop. Extreme cooling in winter causes the air to become cold and heavy: the cold air sinks to give high pressure. Asia, being a much larger landmass, develops much more marked pressure patterns than North America. The annual change in pressure creates very significant winds called the **monsoons** over south and east Asia.

16 From Figs. 4.9 and 4.10 describe carefully, with the aid of a sketch map, the pressure changes from summer to winter and the paths taken by the monsoons in south and east Asia.

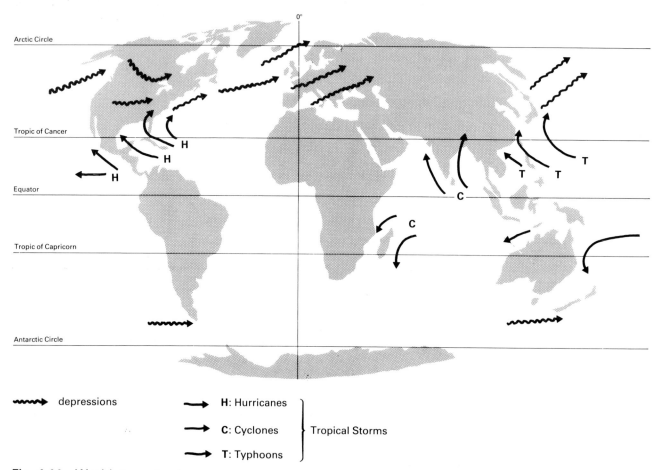

Fig. **4.11** World storm tracks

World Climates, Vegetation and Soils

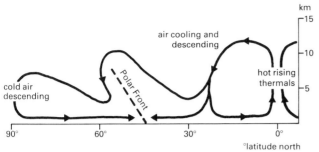

E L P: Equatorial Low Pressure
N E T: North East Trades
S T H P: Sub-tropical High Pressure
W: Westerlies
P E: Polar Easterlies
P H P: Polar High Pressure
D: Depressions

Fig. **4.12** Air motion (horizontal and vertical) in the northern hemisphere

17 Make a copy of Fig. 4.12 and mark on it, in the correct places, the code letters listed in the key.

As the Sun moves north and south from Tropic to Tropic, all the pressure and wind belts move north and south with it. Therefore areas that lie on or near the boundary of two types of pressure belt will experience first one belt and then the other as they move back and forth. Table 4.2, columns I, II and III, summarises these movements.

The fundamental cause of the pressure and wind patterns around the globe is the Sun. The intense insolation in equatorial regions creates instability and low pressure. As the air rises, it cools and then descends to create the subtropical high pressure cells with their outward blowing **Trade Winds**. Fig. 4.12 shows the air motion through the atmosphere both horizontally and vertically in the northern hemisphere.

Distribution of Rainfall

The global pattern of rainfall is closely related to the wind and pressure patterns, and is shown in Figs. 4.13 and 14. In general, a region will experience rainfall if:

a) the air pressure is low (convectional rainfall),
b) tropical storms or depressions occur (cyclonic and frontal rainfall),
c) winds are onshore, especially where the wind has to rise over high ground (orographic rainfall).

I Location	II Summer Pressure & Winds	III Winter Pressure & Winds	IV Rainfall Regime & Type	V Regional Example
0°–6° N & S	Low pressure (Doldrums)	Low pressure (Doldrums)		
7°–15° N & S				
West coast	Low pressure (Doldrums)	Trade Winds		
Continental	Low pressure (Doldrums)	Trade Winds		
East coast*	Low pressure (Doldrums)	Trade Winds		
16°–30° N & S				
West coast	Trade Winds	High pressure		
Continental	Low pressure	High pressure		
East coast*	Trade Winds	Trade Winds		
31°–40° N & S				
West coast	High pressure	Westerlies		
Continental	Low pressure	High pressure		
East coast*	Westerlies	High pressure/Westerlies		
41°–60° N & S				
West coast	Westerlies	Westerlies		
Continental	Westerlies	High pressure		
East coast	Westerlies	Westerlies		

* In India, S. and E. Asia and N. Australia monsoons occur in each season.

Table **4.2** World pressure and precipitation

74 World Climates, Vegetation and Soils

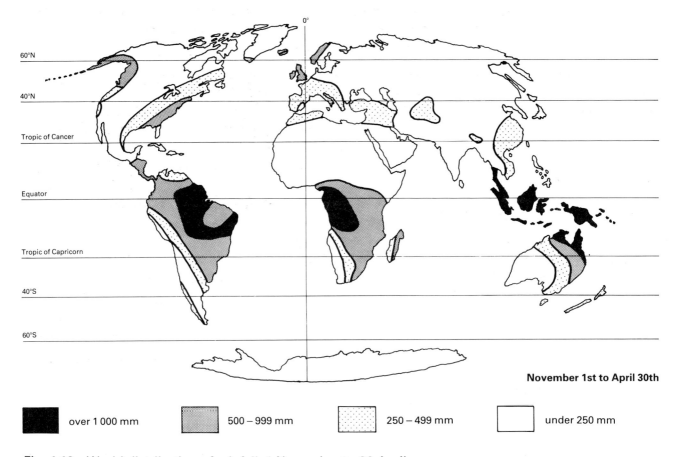

Fig. **4.13** World distribution of rainfall: 1 November to 30 April

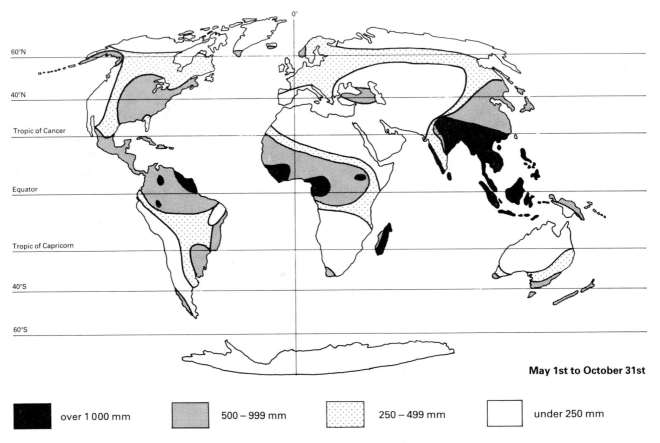

Fig. **4.14** World distribution of rainfall: 1 May to 31 October

World Climates, Vegetation and Soils

Dry regions are those with high pressure, or those that are a long way from a supply of moisture, i.e. the sea. (The hot deserts of the world like the Sahara combine both these features throughout the year.) High latitudes tend to be dry because the cold air can hold little water vapour. The pattern of rainfall which a place experiences during the year is known as its **rainfall regime**.

18 Study Table 4.2 and refer to the maps of world pressure, wind and rainfall distribution in this chapter. Copy the table into your exercise book, including the empty columns IV and V. For each location listed in column I write in column IV a) whether there is summer rain, winter rain, rain all year, or no rain at all; and b) whether the rain is convectional, cyclonic, frontal, or brought by onshore winds.

Check your answers by looking at Fig. 4.15 which shows the rainfall distribution for a model northern continent.

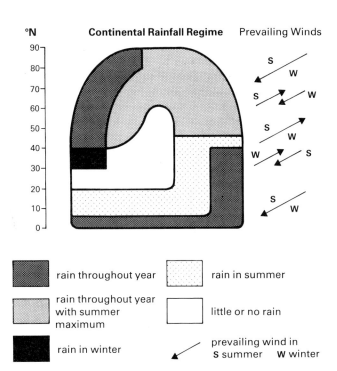

Fig. **4.15** Rainfall regimes on a model continent

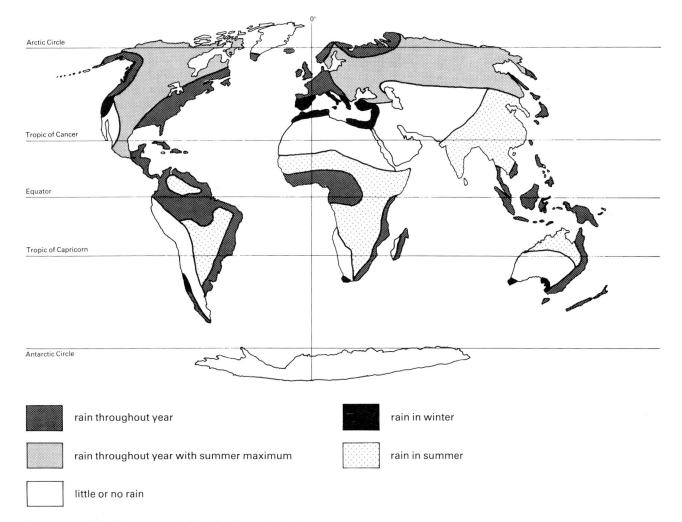

Fig. **4.16** World seasonal distribution of rainfall

> **19** In column V of Table 4.2 write the name of a region to serve as an example for each location listed in column I. Use Figs. 4.13 and 4.14 as well as an atlas.

It should be noted that continental interiors that have rainfall throughout the year have a *summer* maximum, and western coasts that have rain throughout the year have a *winter* maximum.

We have now seen that many regions of the world experience a dry season or have no significant rain at all during the year. The dryness or aridity of a region, however, depends not only on rainfall totals but also on the temperature. Many geographers regard 500 mm of rain a year as the boundary of deserts in hot regions, but only 250 mm a year in cold regions.

> **20** Suggest reasons why this should be so.

Not only is the model continent in Fig. 4.15 a very generalised picture, but so is the world map in Fig. 4.16. Reality is very much more complex, with considerable variations in rainfall from year to year. In the mid 1970s, the monsoon rains of northern India failed to arrive; Britain suffered serious drought in 1975–76, whilst Alice Springs in the Australian desert experienced flooding. Generally speaking, the lower the annual rainfall, the more unreliable the rainfall becomes.

Classification of World Climates

Fig. 4.17 shows the distribution of climatic types around the world, based upon temperature and rainfall

Station		J	F	M	A	M	J	J	A	S	O	N	D	Year
Singapore	°C	25.8	25.7	26.5	26.9	27.2	26.9	27.2	26.8	27.1	26.9	26.3	26.0	26.6
1°N	mm	105	45	105	80	95	95	55	65	135	160	145	95	1200
Kandi		25.0	28.0	30.8	31.7	30.9	28.0	26.4	25.4	26.2	27.9	26.9	25.1	27.7
11°N		0	1	5	30	70	145	190	335	215	55	5	0	1050
Buenos Aires		23.0	23.0	21.0	17.0	13.0	9.0	10.0	11.0	13.0	15.0	19.0	22.0	16.0
34°S		79	71	109	89	76	61	56	61	79	86	84	99	950
Alice Springs		27.9	26.5	24.3	19.5	14.6	11.2	12.4	15.4	18.3	22.1	25.0	26.7	20.3
24°S		20	90	30	5	15	3	10	3	3	25	25	57	286
Rome		6.3	7.7	10.7	14.0	18.5	23.2	26.0	25.0	22.6	16.9	12.1	8.4	16.0
42°N		65	75	35	30	45	20	8	12	50	90	130	80	630
London		3.9	3.8	6.2	8.7	10.7	14.9	17.5	16.2	14.2	10.4	7.1	4.9	9.9
51°N		50	40	40	40	50	45	60	65	50	50	60	50	600
Berlin		−1.8	0.0	4.0	9.0	14.2	17.0	18.9	18.1	14.8	8.8	4.7	0.6	9.0
52°N		50	35	30	55	50	60	90	70	55	55	50	40	640
Astrakhan		−7.0	−7.0	−0.3	10.0	17.8	23.0	25.5	24.0	17.2	9.7	2.0	−4.0	9.2
46°N		9	12	12	10	22	17	17	13	18	20	13	12	175
Moscow		−10.0	−8.2	−4.8	3.5	10.1	15.6	18.5	16.1	11.1	4.2	−2.7	−7.8	3.8
56°N		25	25	35	30	50	75	80	90	55	45	35	35	580
East Port		−4.7	−5.0	−0.5	3.7	8.3	12.3	15.0	14.5	13.0	9.0	3.8	−3.1	5.5
43°N		70	60	70	55	55	70	70	75	80	60	85	70	820
Coppermine		−27.3	−29.1	−24.0	−17.1	−5.1	3.5	8.9	8.0	2.9	−5.1	−14.0	−25.6	−10.0
68°N		20	10	20	15	30	20	40	40	35	35	20	12	277
Rangoon		25.8	26.9	28.5	30.3	29.1	27.5	27.0	27.0	27.6	27.5	27.5	25.7	27.5
17°N		1	5	5	75	310	545	605	515	450	220	80	25	2836
Santis		−8.8	−8.5	−5.8	−2.9	0.7	3.8	5.9	6.0	3.8	0.0	−4.8	−7.7	−1.5
46°N		215	200	180	175	170	245	320	295	230	170	275	190	2665

Table **4.3** Mean monthly temperatures and rainfall for selected stations

World Climates, Vegetation and Soils

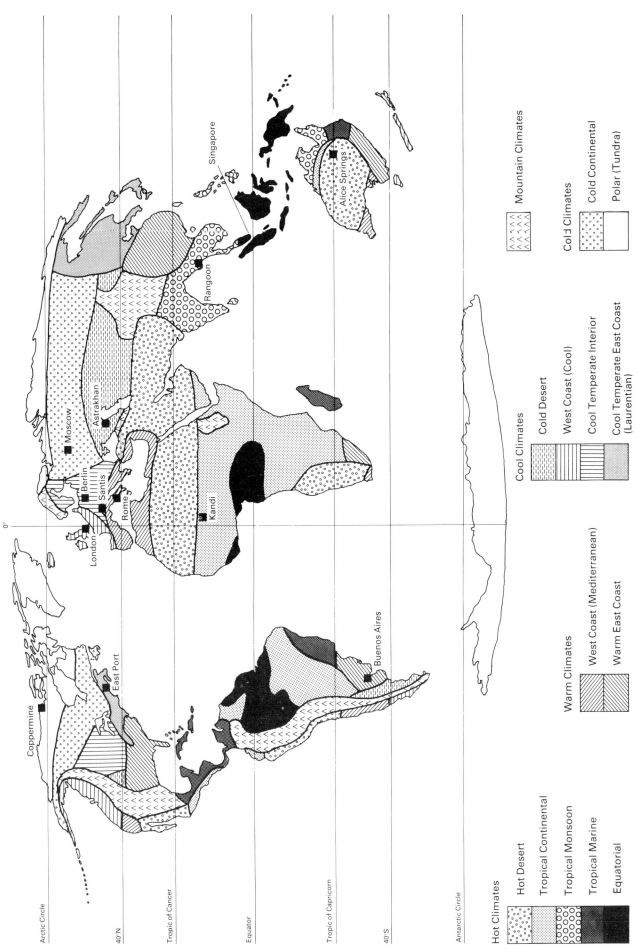

Fig. 4.17 Simplified classification of world climates

characteristics. Although boundaries have been drawn between the regions, it must be remembered that one climatic region merges gradually into the next. Table 4.3 gives mean rainfall and temperature data for each of the representative stations marked on Fig. 4.17. This data can be shown in the form of a graph, as in Fig. 4.18, the graph for London.

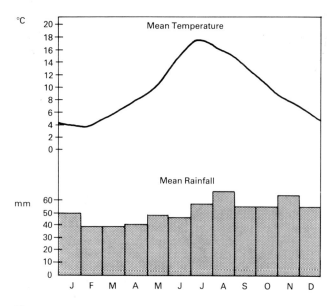

Fig. **4.18** Climatic graph of London

When describing a climate, the following questions should be used as guidelines:

a) What are the mean maximum and mean minimum temperatures during the year? In which month does each occur?
b) What is the annual temperature range?
c) Is the region likely to experience frost? (Frost can be expected in months with a mean temperature below 10°C.)
d) How many months are below 6°C (the minimum temperature at which plants will grow) and how many below 0°C?
e) What is the total annual rainfall?
f) Is there a marked dry season (three or more consecutive months with a total below 25 mm)? When?
g) Is there a marked wet season (three or more consecutive months with a total above 25 mm)? When?

The *causes* of the temperature and rainfall characteristics also need to be described.

Natural Vegetation

The term *natural* vegetation applies to the type of vegetation that would develop in a region without human interference. In places like Britain, very little of the vegetation can be described as natural in that sense, whereas in the Amazon Basin, a very large proportion of the vegetation is natural. The climate of a region is the chief factor in determining the type of vegetation that develops there. Fig. 4.19 shows the main natural vegetation zones around the world.

> **21** Using the data in Table 4.3 and Figs. 4.17 and 4.19, write down the type of vegetation to be found in each climatic region.

In order to grow, plants need adequate light, heat, moisture and nutrients. A plant and the soil it grows in can be regarded as a system, as shown in Fig. 4.20.

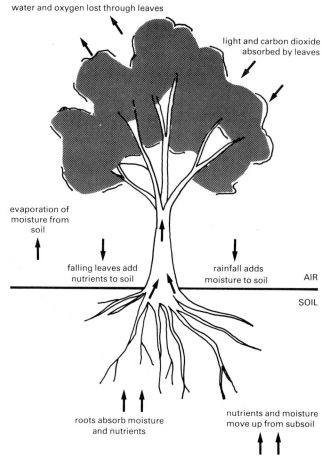

Fig. **4.20** The plant–soil system

World Climates, Vegetation and Soils

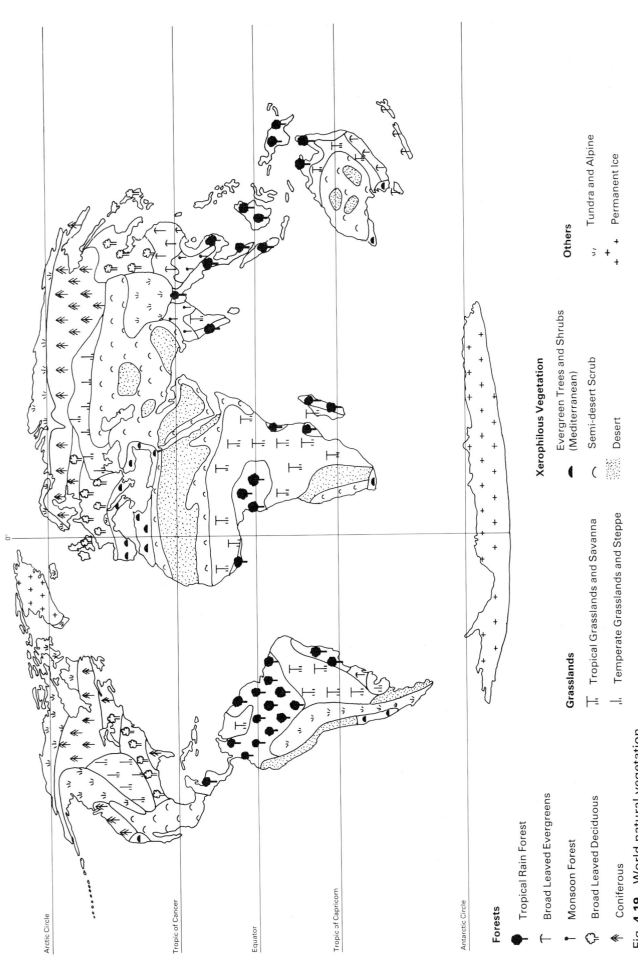

Fig. 4.19 World natural vegetation

> **22** List the inputs and the outputs of the system shown in Fig. 4.20. (You may need to refer back to p. 28.)

In many regions there is a shortage of certain inputs. The plants then have to adapt to those deficiencies. Lack of water can be overcome by one or more of the following adaptations:

a) the roots spreading widely just below the surface;
b) a tap root penetrating deeply to the water table;
c) the leaves being waxy or thorn-like to reduce moisture loss (transpiration) from them;
d) losing leaves during a dry season;
e) storing water in the tissues (as with cacti).

When the temperature falls below 6°C, plant growth ceases and vegetation remains dormant. Cold conditions can be tolerated by:

a) having needle-shaped leaves that reduce the surface area exposed to the cold air (as with **coniferous** trees); or
b) losing leaves altogether during the winter (**deciduous** trees).

For several months of the year many regions have climatic conditions that do not permit plant growth. Therefore plant growth is said to be seasonal. The natural vegetation that develops in a particular climate is known as the **climatic climax vegetation**.

Soils

The nutrients that a plant requires are absorbed through the roots which penetrate the soil. Soil is a complex mixture of **organic** and **mineral** matter. The organic material comes from decaying plant and animal life. The minerals come from the underlying rock as it weathers and decays. Fig. 4.21 shows the soil as a system with inputs and outputs.

> **23** Study Fig. 4.21 and list the inputs and outputs of the soil system.

The three basic mineral materials which form a soil are clay, sand and calcium carbonate. Clay is very fertile, but it drains badly, making it heavy and wet. In hot, dry conditions it becomes very hard. Sand is infertile, but it drains well and warms up quickly after a cold season. Mixed with clay it forms **loam**. Calcium carbonate is fertile and drains well. Mixed with clay it forms **marl**.

> **24** What properties of drainage and fertility do you think loam and marl have?

Clay is particularly fertile because each particle attracts decaying vegetable matter, known as **humus**. The nutrient value of the humus depends largely upon the type of vegetation growing in the soil. Tropical rain forests produce a rich humus: the warm moist climate encourages the decay of fallen leaves. On the other hand, coniferous forests produce a poor humus: the hard, needle-shaped leaves break down only very slowly in the cold climate. The movement of nutrients from the soil into the plant and back again to the soil when the leaves fall is known as the **nutrient cycle**.

> **25** What will happen to the fertility of the soil if the vegetation is cleared and the nutrient cycle broken?

The cross-section of a soil, from the surface down to the bedrock, is called a **soil profile**. A model profile is shown in Fig. 4.22. Each layer in the profile is called a **horizon**.

> **26** Study the soil profile in Fig. 4.22. In what ways does the A horizon differ from the B horizon?

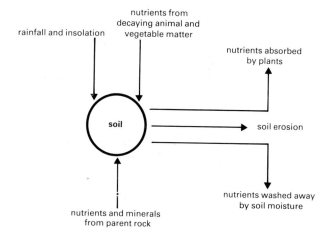

Fig. **4.21** The soil system

World Climates, Vegetation and Soils

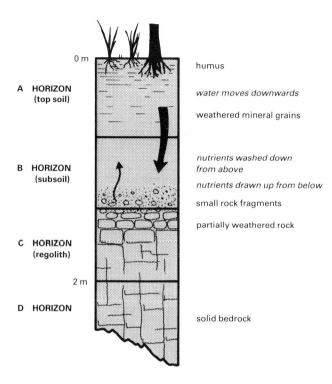

Fig. 4.22 A model soil profile

In regions where evaporation is *less than* rainfall water movement in the soil is downwards. As it moves downwards through the soil, it carries with it minerals and nutrients: this is known as **leaching**. In cool climates, especially in coniferous forest regions, oxides of aluminium and iron are leached downwards to be deposited in the B horizon as an **ironpan** or **hardpan** layer. This restricts drainage on the surface, creating waterlogged conditions. Silica (sand) remains in the A horizon. Such a soil is called a **podzol**. In the hotter conditions of the Tropics, it is the silica that gets leached. The iron and aluminium remain in the A horizon. Such a soil is called **laterite** (so-called because it bakes hard in the sun and can be used for making bricks: *later* is Latin for brick).

In regions where the rate of evaporation *exceeds* the rainfall, water moves up into the A horizon from the B horizon, taking minerals with it. For this reason, many desert soils are potentially very fertile. In the grasslands of the American Prairies and the Russian Steppes, the dense root network of the grass creates a thick, rich humus. When combined with the minerals brought up from below, the most fertile soils in the world are formed: the **chernozems**. Fig. 4.23 shows the profiles of five soil types.

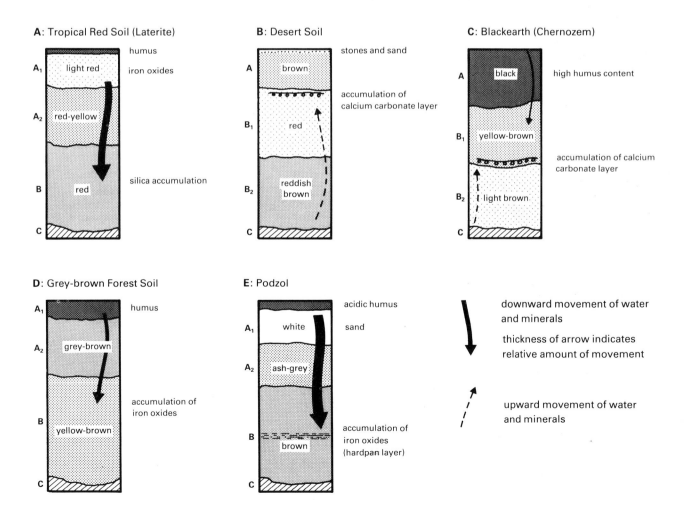

Fig. 4.23 Examples of soil profiles

Fig. 4.25 Vegetation types

World Climates, Vegetation and Soils

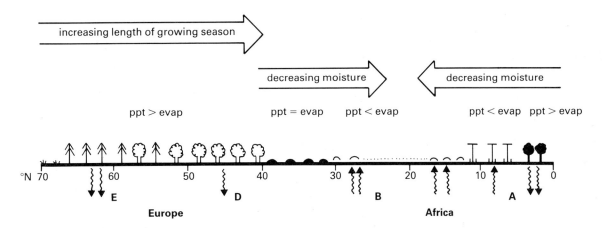

Fig. 4.24 Vegetation types along longitude 20°E, north of the Equator

Fig. 4.24 shows how natural vegetation and certain characteristics of climate and soil change from the Equator northwards to the polar regions along approximately longitude 20°E. Note that sea areas, such as the Mediterranean, have been ignored.

27 a) Describe and give reasons why the soil moisture movement changes along the line of the transect in Fig. 4.24.
b) Between approximately what lines of latitude on the transect would you expect to find i) podzols, ii) laterites? Why would you not expect to find chernozems along this transect?
c) What length of growing season would you find at the following lines of latitude along the transect: 70°, 50°, 30° and 10°? (Refer to Fig. 4.17 and Table 4.3.)

28 Each of the photographs in Fig. 4.25 is of one of the vegetation types indicated on Fig. 4.24. Describe the vegetation in each of the photographs and attempt to identify its type.

Soil Erosion

People have influenced natural vegetation and soils to a considerable degree in many parts of the world. By removing the vegetation cover for agriculture and building they have broken the nutrient cycle and encouraged **soil erosion**. The three major types of soil erosion are:

a) **sheet wash**, where large areas of topsoil are washed away in rain storms;

b) **gullying**, where water erodes deep channels or gullies;
c) **wind erosion**, where dry topsoil is blown away.

29 Study the photographs in Fig. 4.26. Describe the land surface in each case, and say which type of erosion each photograph depicts.

The major causes of soil erosion are listed below:
a) **Overcropping**: growing a crop year after year depletes the soil of nutrients and breaks down its natural structure. This is particularly risky in regions where the mean annual rainfall is only just sufficient for cropping: a year of drought can cause crop failure and initiate erosion by wind. This occurred in the famous 'Dust Bowl' of the plains of the Mid-West USA in the 1930s. An area twice the size of Britain suffered severe soil erosion, from Nebraska in the north to Texas in the south. Dry winds carried topsoil and dust thousands of kilometres.

The soil can be nursed back to life by planting special grasses and not planting crops for several years. Wind erosion can be prevented by planting shelter belts of trees around the fields.

b) **Overgrazing**: grasslands in dry regions can be removed by grazing too many animals. Photograph B in Fig. 4.26 shows erosion after overgrazing in Kenya. In West Africa, overgrazing is one of the causes of the southward spread of the Sahara Desert.

Restricting the size of herds would help to minimise soil erosion, but this is not easy as cows are considered as sources of wealth.

c) **Tree felling** and ploughing of steep slopes can cause the soil to be washed down the slope. One hour's heavy rain on a slope of 1 in 24 (quite a gentle slope) can remove 20 kg of soil per hectare.

One answer is to terrace the hillsides, as is done in many parts of S.E. Asia. On reasonably gentle slopes, sheet wash can be avoided by ploughing the land parallel to the contours (**contour ploughing**) and planting the crops in strips.

30 Fig. 4.27 A and B shows two methods that have been used to combat soil erosion. For each photograph, state the type of soil erosion you think is being prevented, and describe the prevention method being used.

Fig. **4.26** Examples of soil erosion

World Climates, Vegetation and Soils

Fig. **4.27** Prevention of soil erosion

The Natural Regions of the World

Now that we have studied how the climate, vegetation and soil systems work, we shall be looking at the way they operate together in different natural regions of the world. The regions that we shall study are based upon the climatic regions shown in Fig. 4.17. It must be remembered that there are no definite boundaries to these regions: one merges gradually into the next.

> **31** As each region is dealt with in turn, describe the climate using the guidelines set out on p. 78 and draw a climatic graph like the one shown for London on p. 78. You should also study the soil profiles in Fig. 4.23.

Equatorial Regions

> **32** Study the distribution of these regions on Fig. 4.17. Give a suitable geographical name to each area which has an Equatorial type climate.

The natural vegetation of these areas is dense **evergreen forest**, called **tropical rain forest**. A picture of such a forest can be seen in Fig. 4.25B. Despite the chaotic appearance of the forest, it does have a basic and simple structure, as shown in Fig. 4.28.

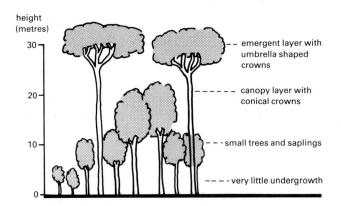

Fig. 4.28 The structure of tropical rain forest

35 Find out which regions produce the following plantation crops: a) rubber, b) cacao, c) oil palm.

One major project of agricultural development is taking place at Jari in Brazil, where rice is being grown in the rich alluvium of the Amazon valley. Trees suitable for the production of wood pulp are being planted in place of the natural forest. The Jari project is just one example of the development of the Amazon Basin which is now taking place. Some scientists fear that wholesale clearance of this forest could result in a considerable reduction in the region's rainfall as well as a serious reduction in the oxygen level of the world's atmosphere (since vegetation gives out oxygen through its leaves).

36 Refer back to Fig. 4.20 and explain why scientists have these fears.

On the other hand, some scientists say that the oxygen is in a closed cycle rather like that of the nutrient cycle and so does not contribute to the world's supply.

33 Describe the structure of the tropical rain forest as shown in Fig. 4.28.

34 The trees of the forest are evergreens rather than deciduous: can you explain why?

37 Locate Jari (400 km west of Belem) in your atlas. How far is it from Brasilia and Rio de Janeiro? (This illustrates how far it is from the major centres of population.)

38 Approximately what proportion of Brazil is covered by rain forest?

There is a very large range of tree species in the rain forest: some areas have over 200 tree species per hectare. Mahogany, teak and ebony are amongst those with the greatest commercial value. However, getting them out of the forest is difficult because they are scattered over a very large area. Tropical red soils (or laterites) predominate. If the trees are felled, the nutrient cycle is broken and severe leaching and erosion can occur.

The loss of tropical rain forest over recent years has been considerable, and this loss is accelerating. The traditional ways of life, hunting and gathering and shifting agriculture, tended to work in harmony with the forest system. Today, trees are being cleared to provide not only more agricultural land for the rapidly growing populations, but also for their use as fuel (over half the human population uses timber as a fuel). It has been estimated that a total area equivalent to the size of Switzerland is being cleared every five years.

Replacement of the forest by commercial plantations has gone on since the European colonial expansion in the eighteenth and nineteenth centuries.

Tropical Continental Regions

39 How can you explain the fact that rain only falls for part of the year (the summer) in Tropical Continental regions?

Grasses are able to withstand the seasonal drought. Drought-resistant trees like the acacia and the baobab are also found. This mixture of grasses and trees is called the **Savanna** in Africa, the **Llanos** in Venezuela and the **Campos** in Brazil.

World Climates, Vegetation and Soils

Laterites are the most widespread soils. In the dry season the upper layers are baked hard by the Sun, especially if the vegetation cover has become thin due to overgrazing or overcropping. This encourages erosion in the wet season.

Cattle grazing is a major activity on the grasslands. In order to improve the quality of the grasses, many tribes traditionally burn it at the end of the dry season: this encourages new growth during the wet season. As most trees cannot withstand fire, many authorities believe that, in fact, the grasses are *not* the natural vegetation at all. The climatic climax may be forest, but this has been partially cleared by the fires.

The Masai of East Africa are a nomadic tribe who move from pasture to pasture with their cattle. During the mid-1970s the region's rainfall, usually fairly reliable, ceased to arrive altogether. Thousands of people died of starvation in the Karamoja district of N.E. Uganda and the neighbouring parts of the Sudan. Many were forced to move south, putting an even greater strain on the low productivity of the region's agriculture. Agricultural improvement is difficult due to the low, unreliable rainfall and the lack of capital investment. Commercial agriculture, with millet, maize and beans is important near the towns. Tobacco is a major export crop for Zimbabwe. On the Llanos and Campos, commercial cattle rearing is important.

In Africa and northern Australia, the Tropical Continental region can be regarded as a transition zone, or zone of gradual change, from tropical rain forest to semi-desert and desert. In Africa, the Sahara is gradually encroaching southwards as a result of decreasing rainfall and overgrazing. This region (on the desert margin) is known as the Sahel.

> **40** Refer back to p. 83 and explain why overgrazing can help the desert to advance.

Hot Deserts

> **41 a)** On an outline map of the world, shade and name the following hot deserts: Arabian, Arizona, Atacama, Australian, Kalahari, Sahara.
> **b)** Study Figs. 4.9 and 4.10 again to remind yourself of the reasons why these regions are dry.

Rainfall is not unknown in the desert. One year's supply however can fall in one brief storm, resulting in very high run-off rates (see Chapter 8).

The mean annual temperature statistics do not reveal the very high *diurnal* range that often occurs. In the hot season, maximum daytime temperatures can reach 49°C and fall to a minimum of 15°C at night. In the cool season, ground frosts are not unknown.

> **42** Explain why there should be such high diurnal ranges.

Plant life is very sparse but nevertheless does exist. It must be able to withstand very long periods of drought: such plant life is said to be **xerophytic**.

> **43** What methods can xerophytic vegetation use to survive lack of rainfall? (Re-read p. 80.)

Some species can remain dormant for years until a brief shower enables them to complete a full life-cycle during the duration of the rain.

Indigenous peoples of the deserts have highly specialised life-styles to survive the extreme conditions. Probably the most famous are the Bedouin: the nomadic herdsmen of the northern Sahara and Arabia.

Settled agriculture is possible at the **oases** where underground water supplies are tapped. River valleys such as the Nile, Tigris and Euphrates have long been used to support agriculture by irrigation. Large scale commercial agriculture has been made possible where massive amounts of capital have been invested to provide irrigation over large areas, such as the schemes in Israel, the south-western USA and Soviet Central Asia.

> **44** Why are desert soils potentially very fertile? (See Fig. 4.23.)

The prolonged and intense insolation received by desert regions could be harnessed for the production of electricity. One problem however is the long distance of most desert regions from major markets.

Cold deserts are defined as those deserts that have at least one month with a mean temperature below 6°C. These regions are colder in winter because of their greater latitudes. The soils, vegetation and summer climate are very similar to those of the hot deserts.

Tropical Monsoon Regions

> 45 Locate and name these regions from Fig. 4.17 and your atlas.
>
> 46 What are monsoons, and why do they occur? (Read again p. 72 and study Figs. 4.9 and 4.10.)
>
> 47 How does the change in wind direction affect the rainfall distribution a) throughout the year and b) across S.E. Asia and northern Australia?
>
> 48 With the aid of a relief map of S.E. Asia in your atlas, describe the effect that high ground has on the rainfall. Referring back to p. 56, explain the effect that you have just described.

Cherrapunji, 1300 m up in the Khasi Hills in Assam, India, is one of the wettest places on the Earth: its mean total rainfall for the year is 10 871 mm.

In India, Bangladesh and Burma, three seasons can be identified: a) the cool dry season, b) the hot dry season and c) the hot wet season.

> 49 Using the data for Rangoon in Table 4.3, identify and name the months of the year in which each of these seasons is found.

This region is particularly prone to tropical cyclones, which on the north coast of Australia are called 'Willy Willies'.

Wet-rice agriculture is a very important subsistence activity in the river valleys. Dry-rice cultivation, with wheat, sorghum and millet, is practised in drier areas. Tea and cotton are important commercial crops. The unreliability of the monsoon, especially in central and N.W. India, coupled with a rapidly rising population, has caused famine to occur with increasing frequency. The 'Green Revolution' has increased rice yields considerably through the introduction of new rice varieties and fertilisers. Improved irrigation techniques have been established in the drier zones.

Warm West Coasts (Mediterranean Type Areas)

> 50 Identify from Fig. 4.17 the regions of the world which have a Mediterranean type climate. Give each of these regions a suitable geographical name.
>
> 51 How can you account for the marked seasonal nature of the rainfall distribution in these regions?

The rainfall can be very unreliable. Six of the twelve months of the year can be completely dry in some years. The rain can also be very heavy, especially in autumn and spring.

As the regions are on the border of tropical and temperate zones, very contrasting **air masses** can be experienced. An air mass is a mass of air that has particular properties of temperature and humidity. Cold air masses from the poleward side can bring frosts in winter. In summer, hot dusty conditions can be brought from the deserts on the equatorial side.

> 52 Refer back to Fig. 3.39. What are the specific local winds that bring cold and hot air masses to the Mediterranean region of Europe?

The vegetation is adapted to the summer drought. Thorny trees and shrubs such as olive and gorse cover large areas. Such areas are called the **Maquis** in France and the **Chaparral** in California. Trees include pines and evergreen oaks.

Over the past few thousand years, people have been cultivating the Mediterranean regions of Europe and North Africa. This has resulted in the clearance of the natural vegetation and severe soil erosion. The Maquis may in fact be scrubby secondary growth after forest clearance by farmers, who subsequently exhausted the goodness of the soil. Although a number of farmers are poor and have to work hard for a meagre income from their land (especially in Europe and North Africa), intensive commercial farming is important. Oranges, peaches, apricots, figs, dates and olives are the main crops, with California a leading fruit exporter. Vines are widespread. The Common Market Agricultural Policy is a major instrument in increasing the agricultural output of the Mediterranean areas of France, Italy and Greece.

The almost certain occurrence of hot summer sunshine has made southern Europe a flourishing tourist

World Climates, Vegetation and Soils

area, especially since cheap jet travel has been available from the cooler areas further north. The growth of the Hollywood film industry is partly due to the fact that filming can be done out of doors for most of the year.

> **53** Study Fig. 4.29 and do the following:
> a) On a copy or tracing of the map, shade all areas with under 1008 mb pressure in one colour and all areas with over 1012 mb in another colour.
> b) On the same outline, draw on the position of the 40°C and 25°C isotherms.
> c) Compare the weather conditions at stations A, B and C.
> d) To what extent does the information you have studied in a) to c) conform to what you have learnt about the climate of these regions?

Cool Interior and West Coast Regions (Cool Temperate Regions)

> **54** Study Fig. 4.17 and list the regions of the world that come into these categories.
>
> **55** How can you account for the differences in temperatures in summer and winter between London and Berlin (Table 4.3)?

The climatic data for these regions conceal great variations that occur from year to year and within each season. In order to study these variations we shall concentrate on the European region. Fig. 4.30 shows the daily maximum and minimum temperatures for the winter months of 1966–67 and 1962–63 in Sheffield.

> **56** From Fig. 4.30 estimate for each winter shown:
> a) the proportion of the time the daily maximum was at freezing point or below,
> b) the proportion of the time the daily minimum was at freezing point or below,
> c) the mean temperature.

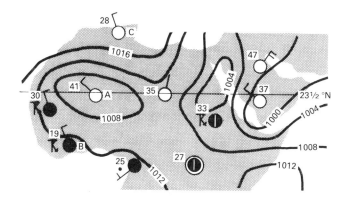

Fig. **4.29** Synoptic chart for North Africa, 1200 hrs G.M.T. 14 July 1967

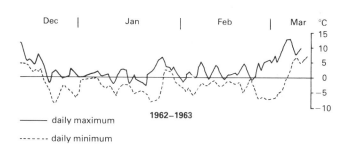

Fig. **4.30** Daily maximum and minimum temperatures at Sheffield

Warm East Coasts

Monsoon winds and tropical storms also occur on the warm east coasts of North America and China. However, it is cooler and drier than in the Tropical Monsoon regions. In China wet- and dry-rice cultivation are very important. In the USA and Australia commercial farming dominates, with tobacco, cotton, maize and cattle. Sudden bursts of cold air from higher latitudes in the winter and hot dry air from the continental interiors in the summer can damage crops.

The reason for the great variability of the weather over Europe as a whole is the occurrence of different and very contrasting air masses which can affect the

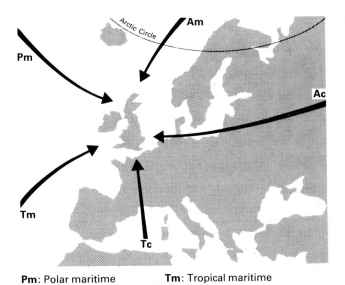

Pm: Polar maritime **Tm**: Tropical maritime
Am: Arctic maritime **Tc**: Tropical continental
Ac: Arctic continental

Fig. **4.31** Air masses which occur over Europe

continent. Fig. 4.31 shows these air masses. Table 4.4 summarises the weather conditions brought by each type.

In general the cold air masses (Polar maritime, Arctic maritime and Arctic continental in winter) will be *unstable* because the lower layers of the air will warm up on the journey to Europe. The warm air masses (Tropical maritime, Tropical continental and Arctic continental in summer) will be *stable* because the lower layers will cool down as they come towards the continent (see p. 54).

57 Why should the Arctic continental air mass be cold in winter but hot in summer?

Fig. 4.32 is the synoptic chart for 11 Dec. 1981. It shows a particularly cold Arctic maritime air mass affecting Britain. In this particular case, the air mass was associated with a depression that moved east across southern England, bringing heavy snowfalls. About 200 mm fell during the day, the worst pre-Christmas snowfall over S.E. England since 1950.

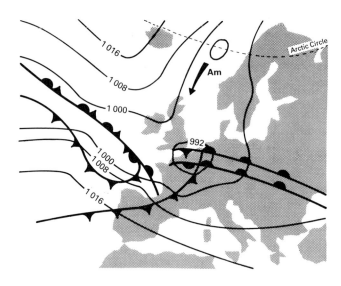

Fig. **4.32** Synoptic chart for Western Europe, 1200 hrs G.M.T. 11 December 1981 – a particularly cold spell of Arctic maritime air

Air mass	Source region	Summer conditions	Winter conditions
Polar maritime	N. Atlantic	cool, showery	cold, showery, some snow and frost
Tropical maritime	Azores	very warm, coastal fog and cloud; sunny and humid inland	mild, humid and cloudy
Arctic maritime	Arctic	cold, showery	very cold, snow showers and frost
Arctic continental (or Polar continental)	Russia	hot, dry and sunny	very cold and frosty, sunny
Tropical continental	N. Africa	hot, dry and hazy	very mild, dry

Table **4.4** Air masses and their characteristics over Europe

World Climates, Vegetation and Soils

> **58** What particular aspect of the pressure pattern shown on Fig. 4.32 was the cause of the Am air reaching Britain?
>
> **59** Refer back to Fig. 3.38 and draw sketch maps to show the pressure patterns over and near N.W. Europe that would bring to Britain each of the air masses listed in Table 4.4.

In Table 4.4 the air masses have been listed in order of the frequency with which they occur: the Polar maritime is the most common. The climate of Europe can be said to be determined by the frequency with which the different air masses affect the region.

The climatic climax vegetation of the west coasts and cool interiors is deciduous forest in Europe: the most common trees are the oaks, beeches, elms and ash. However, in Scandinavia coniferous forest is more important because the climate is cooler.

> **60** Read again p. 80 to remind yourself of the definition of deciduous. Why should deciduous trees dominate the natural vegetation in cool temperate climates?

Large areas of the natural forest have now been cleared for both urban and agricultural uses. Table 4.5 shows the approximate percentage breakdown of the major land uses in the British Isles.

Land use	% of total area
Urban land	6.4
Forest land	6.1
Improved farmland	54.7
Unimproved land	25.8
Other land	7.0

Total land area = 30 640 000 hectares

Table **4.5** Land use in the British Isles

Unimproved land includes the areas of upland moorland. Other land includes quarries, reservoirs and military training grounds.

> **61** Work out how many hectares of forest there are in the British Isles. Find out why a lot of this forest cannot be described as climatic climax vegetation (natural vegetation).

There is a great variety of soils to be found. The fertility of the soils depends very much upon the nature of the bedrock, but overall they are very fertile as the humus is replenished each year by the leaf-fall.

The natural soils are the **grey–brown forest soils** but these have been altered considerably over centuries of agriculture. They have been changed most of all over the last thirty years as commercial pressure has demanded an increase in crop yields. The fertility of the soil is no longer maintained by crop rotations. Fields may be farmed continuously under one crop; the fertility is kept up with large amounts of fertilisers. Diseases are kept down with fungicides, and insects kept at bay with insecticides. So many chemicals are now being sprayed onto the fields that serious pollution is beginning to appear as the various compounds accumulate in the soil and rivers.

Cool Temperate East Coast Regions

> **62** Explain why these regions have much colder winters than the west coast regions in the same latitudes. (Hint: the westerlies and ocean currents.)

Precipitation comes from depressions which move north up the coast in winter (bringing heavy snowfalls) and east from the interior in summer. Typhoons occur in Japan.

High population densities exist in many parts of these regions, with some of the world's largest cities.

> **63** Referring to Fig. 4.17 and your atlas, identify the major cities that are found in the east coast regions.

Agriculture tends to be very intensive, serving the large urban markets. In Japan the chief crops are rice and vegetables. In North America mixed farming is based on dairying and market gardening. The contrasts in agriculture between the two regions are based on cultural differences rather than physical differences of climate and soil. In both regions, however, agriculture is limited in a northward direction, and there are many areas where the soils are too thin and stony and slopes too steep for good farming.

> 64 Explain why agriculture becomes more difficult the further north you go in the east coast regions.

Polar and Cold Continental Regions

Antarctica is not included here as it is entirely covered by ice. The natural vegetation of these regions is coniferous forest, giving way northwards to tundra.

> 65 How are coniferous forest trees adapted to survive the very cold climate? (See p. 80.)
>
> 66 Why are polar and cold continental regions not represented in the southern hemisphere?

The huge stands of pine, spruce, fir and larch (which is a deciduous conifer) have only evolved since the retreat of the glaciers at the end of the Pleistocene Ice Age.

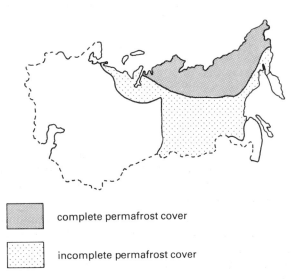

complete permafrost cover

incomplete permafrost cover

Fig. **4.33** Distribution of permafrost in the USSR

One species will cover hundreds of square kilometres: this makes commercial exploitation much easier than in the tropical rain forests. The coniferous forests provide the world with most of its supply of softwoods, principally for wood-pulp, paper and chemicals. The demand is now outstripping the supply: careful forest conservation and management is required to prevent serious depletion of the forests.

To the north, the climate becomes too severe even for coniferous trees. They gradually give way to the mosses, lichens and dwarf flowering shrubs which cover the **tundra** regions. Both climate and soils make agriculture very difficult.

> 67 After studying the climate figures for Moscow and Coppermine in Table 4.3 and reading the description of podzols on p. 81, list the physical difficulties which face farmers in cold continental regions.

Fig. 4.33 shows the distribution of **permafrost** in the USSR. Permafrost is permanently frozen ground. The top few centimetres thaw out for a brief time in the summer.

> 68 From Fig. 4.33 describe the distribution of permafrost in the USSR.

You will learn more about permafrost areas in Chapter 6. It impedes all human activity. Houses and roads need special foundations, and mining, excavation, oil drilling and pipeline laying are very difficult and expensive. This is particularly significant in Siberia, where there are extremely valuable mineral resources which the Soviet government is beginning to exploit.

Mountain Climates

Although climatic data for Santis in the Swiss Alps have been given to represent mountain climates, it must be pointed out that there is not one climate for mountains. The complex pattern of relief: exposed summits and plateaus, deep sheltered valleys and steep slopes, means that the climatic conditions change greatly over quite small distances.

69 Fig. 4.17, which shows the distribution of mountain climates, provides another reason why there is not one single mountain climate: what is this reason?

As we saw in the last chapter, the climate will become colder and wetter as the altitude increases. There will be much more snowfall on mountains: many mountainous areas rise above the permanent **snow-line**.

70 Write in your own words what you think the permanent snow-line is.

71 From Fig. 4.34 calculate how much higher the snow-line is in the Tropics compared with mid-latitudes. Give reasons for this difference in height.

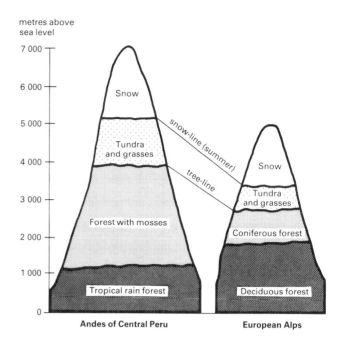

Fig. **4.34** Vegetation zones in mountain regions

The heights given in Fig. 4.34 are for the summer. The snow-line will fall in winter.

72 Study Fig. 4.35.
a) What is the annual change in the height of the snow-line i) on the north-facing slope and ii) on the south-facing slope? Study again Fig. 3.17 and say why this difference exists between the north- and south-facing slopes.
b) Explain why there will be very little change in the height of the snow-line during the year on tropical mountains.

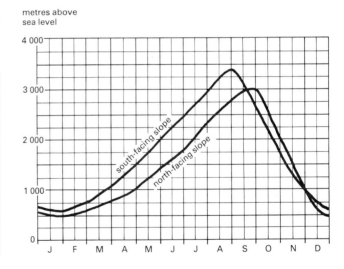

Fig. **4.35** Mean height of snow-line in northern Tyrol, Austria

Another significant line shown on Fig. 4.34 is the **tree-line**, above which it is too cold for tree growth. Only grasses and tundra vegetation are found above the tree-line. In mountains of the mid-latitudes, these grasses make very important summer pastures for sheep, cattle and goats. The annual movement of herds up the mountain from the valley bottom is known as **transhumance**. The floor of the valley is used for the cultivation of fodder crops in the summer months.

73 Read again p. 51 and revise the influence that valleys have on temperatures.

Strong winds can be generated by the descent of *cold* air down the valley: these winds are called **katabatic** winds. During the day, *warm* air rises up the valley to give **anabatic** winds (see Fig. 4.36).

Many mountain regions are isolated and inaccessible. Those that are accessible to wealthy urban populations are rapidly being developed for tourism (especially skiing) and hydro-electricity production. Such developments do involve considerable changes to the landscape and natural environment.

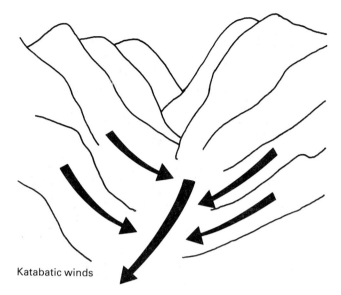

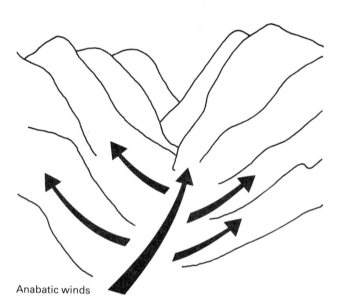

Fig. **4.36** Mountain winds

Fig. 4.37 shows the increase in the amount of CO_2 (measured in parts per million) between 1860 and 1960. Fig. 4.38 and Table 4.6 show some changes in the climate of London over a similar period.

Date	First snowfall	Last snowfall
1841–70	21 Nov	17 Mar
1871–1900	23 Nov	12 Apr
1901–30	25 Nov	15 Apr
1931–60	8 Dec	1 Apr

Table **4.6** Average dates of the first and last snowfalls in London

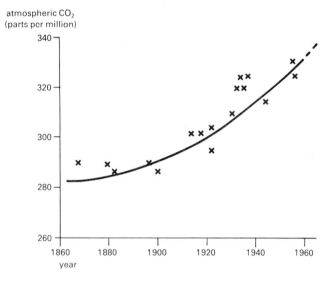

Fig. **4.37** The increasing proportion of carbon dioxide in the atmosphere

74 Is there any correlation between the information shown in Fig. 4.37, Fig. 4.38 and Table 4.6?

Man's Influence on Climate

In the last chapter we saw how settlements can change the local climate by creating 'heat islands'. Settlements and industry consume a great deal of energy. This energy is obtained primarily from the burning of fossil fuels: coal, oil and natural gas. When they burn, carbon dioxide (CO_2) is given off into the atmosphere. As a result, the proportion of CO_2 in the air is increasing.

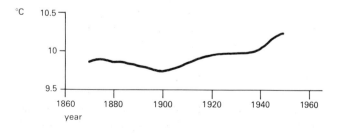

Fig. **4.38** Mean annual temperatures at Greenwich

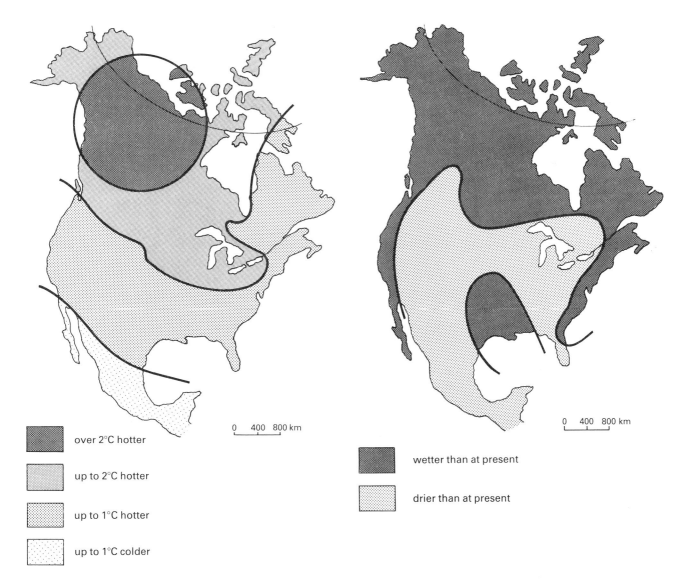

Fig. 4.39 (a) Possible temperature changes by year 2020 in North America

Fig. 4.39 (b) Possible rainfall changes by year 2020 in North America

Carbon dioxide is thought to have a **greenhouse effect** on the atmosphere. It absorbs the outward radiation of heat from the ground, so raising the temperature of the air. Scientists are beginning to get very concerned about the effects that increasing concentrations of carbon dioxide may be having on the global climate. A computer model has estimated that by the year 2050, the global average temperature will have risen by about 2°C. The increase at the Equator would be very little, but at the Poles it would be several degrees more than the average. Figs. 4.39 (a) and (b) and 4.40 (a) and (b) show climatic effects in North America and Europe that may take place by the year 2020.

75 Calculate the expected mean annual temperature in the year 2020 for Coppermine, East Port, London, Berlin and Moscow. (Refer to Fig. 4.17 and Table 4.3.)

The computer forecast is a model based upon our present knowledge of processes in the atmosphere. This knowledge is incomplete for large areas of the globe. Furthermore, scientists are not completely sure about the precise effects increasing amounts of carbon dioxide will have on the atmosphere, especially in relation to other pollutants such as dust. If the model proves correct, there will be serious effects upon the world's food supplies. For example, the maize yields of the USA could drop by 11%.

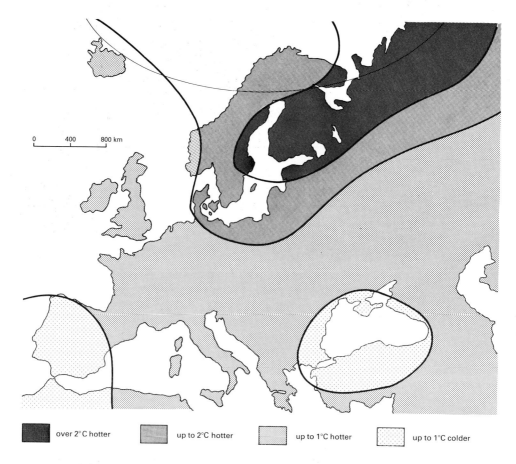

Fig. **4.40 (a)** Possible temperature changes by year 2020 in Europe

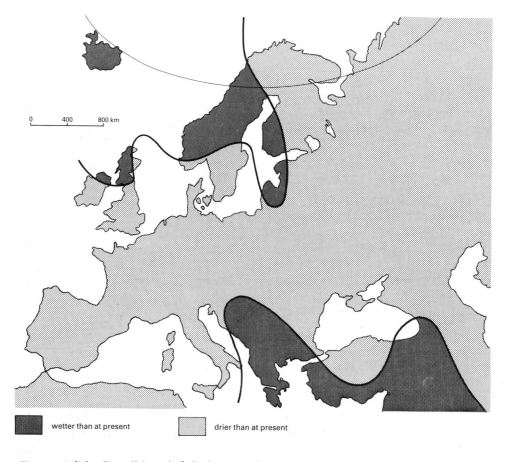

Fig. **4.40 (b)** Possible rainfall changes by year 2020 in Europe

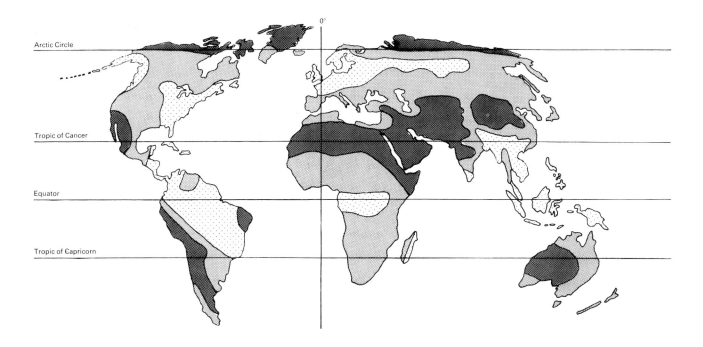

Fig. **4.41** World variability of rainfall

Workshop

1 Study Fig. 4.41. What do you understand by the term 'variability of rainfall'? Test the following hypothesis: *that rainfall variability around the world increases as mean annual rainfall decreases.* (Read pages 73–6 first.)

2 Keep an 'airmass diary' for your home area for at least one month. For each day, record the airmass that is having most influence. (You can find this out by studying the synoptic chart in *The Times, The Daily Telegraph* or *The Guardian*.) Also note down the incidence of depressions, fronts and anticyclones. Record the weather conditions for each day. Can you see any correlation between type of airmass and weather conditions? Attempt an explanation of any correlations you find.

3 Study Fig. 4.42. This shows how the relationship between temperature and humidity affects a European. Using the statistics in Table 4.3, plot (on a tracing of the diagram) the positions of the climatic stations for January and July. The position of London in July has already been done. Remember that you have plotted the mean monthly figures: daily weather conditions may be very different. How do Europeans overcome the discomforts of living outside the comfort zone?

4 Choose any one region in each of the cold, cool, warm and hot climates. For each region you choose: a) describe the natural environment (climate, vegetation and soils), b) describe the ways in which man has changed that environment in the past and at present, and c) discuss what future developments there may be in agriculture in the area over the next 25 years.

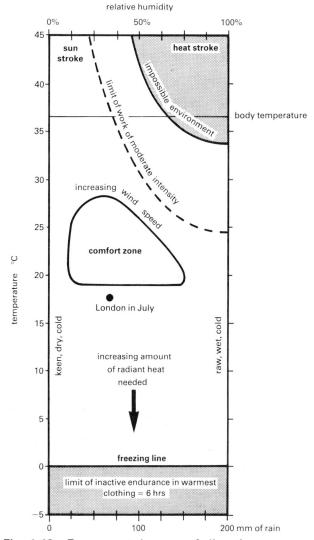

Fig. **4.42** European tolerance of climatic ranges

River Processes and Associated Landforms 5

The landscape in most parts of the world appears to be very stable. None the less it is in a constant state of slow but remorseless change, quite different from some of the abrupt volcanic changes studied in Chapter 1. The science of **geomorphology** is the study of the landforms that develop from these changes, with particular attention being paid to the **processes** that are responsible. (These processes were introduced in Chapter 2.) Wearing down (**denudation**) of the landscape will occur when rock is removed by **weathering** and **erosion**. The carriage of this material downslope is termed **transportation** and the eventual build-up of the eroded material is termed **deposition**. It is this constant redistribution of material over many thousands of years, both on a local and continental scale, that has fashioned today's landscape.

Fig. 5.1 shows the three main zones of the Earth – the air, the land and the sea – which scientists refer to as the atmosphere, lithosphere and hydrosphere respectively. The boundary between each of these zones is known as an **interface**. The diagram shows that the landscape can be divided into a number of sections, each related to a particular 'agent' which happens to be dominant in that location. The major agents are:

a) running water (fluvial)
b) ice and frost (glacial and periglacial)
c) the sea (coastal)
d) wind (aeolian)

Each forms the basis of a separate **geomorphological system**, though there is likely to be a good deal of overlap between systems. The processes of weathering, erosion, transportation and deposition will operate within each system, though in rather different ways; consequently their effects on landform development will vary. It is these differences that will be explored in the following chapters. Finally you will note that volcanic and tectonic systems as well as atmospheric systems are also shown on the diagram; these have been examined in Chapters 1 and 3.

River Processes

Denudation by running water, in the form of streams and rivers, is generally regarded as the 'normal' process of landscape evolution. This is because nearly 70% of the Earth's land surface is drained by river systems, the remaining 30% being covered by ice caps and deserts. However, rivers account for a very small proportion of the Earth's total fresh water – only 0.03%!

> **1 a)** Apart from rivers, where else is fresh water found?
> **b)** Review the 'water cycle' discussed in Chapter 3.

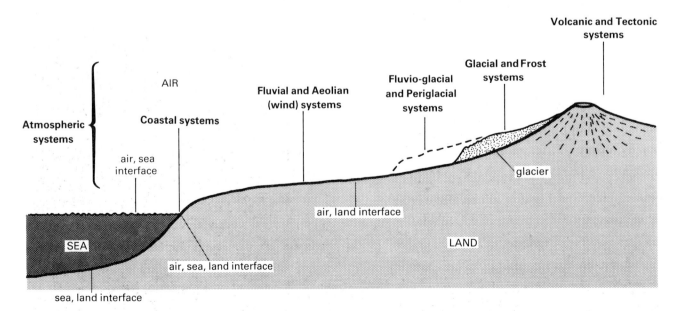

Fig. **5.1** Geomorphological systems

Despite this small percentage, the actual amount of water carried by rivers during an average year is considerable – it is enough to drown the Earth's land surface to a depth of approximately 30 cm.

A glance at Fig. 5.2 provides a good idea of the power of running water and the impact it can have on landform development, though the example chosen is rather exceptional. The River Zambezi has carved out a deep gorge (spanned by the railway bridge), with the eroded material being transported away downstream.

Scientists have estimated that the rivers of the world carry an average of 20 000 000 000 tonnes of eroded material to the sea each year, though this amounts to an effective lowering of the entire land surface of the Earth by only 3 cm every 1000 years. British geomorphologists have also tried to establish the rates of landscape denudation. It was calculated for example that the Cotswold Hills are being lowered by rivers at a rate of 3.16 cm every 1000 years.

> **2** Locate the Cotswolds in your atlas.
> **a)** Find the average height of the hills in metres.
> **b)** Calculate the number of years it would take to lower the hills to present sea level at an average rate of 3.16 cm/1000 years.
> **c)** Find the height of your home above sea level and perform a similar calculation using the same rate of lowering.
> **d)** What other changes are likely to occur during the time you calculate, which will affect the lowering process by rivers?
> **e)** Why might rates of lowering vary between different locations?

You should by now appreciate the great time scale involved in landscape evolution.

Fig. **5.2** The Victoria Falls

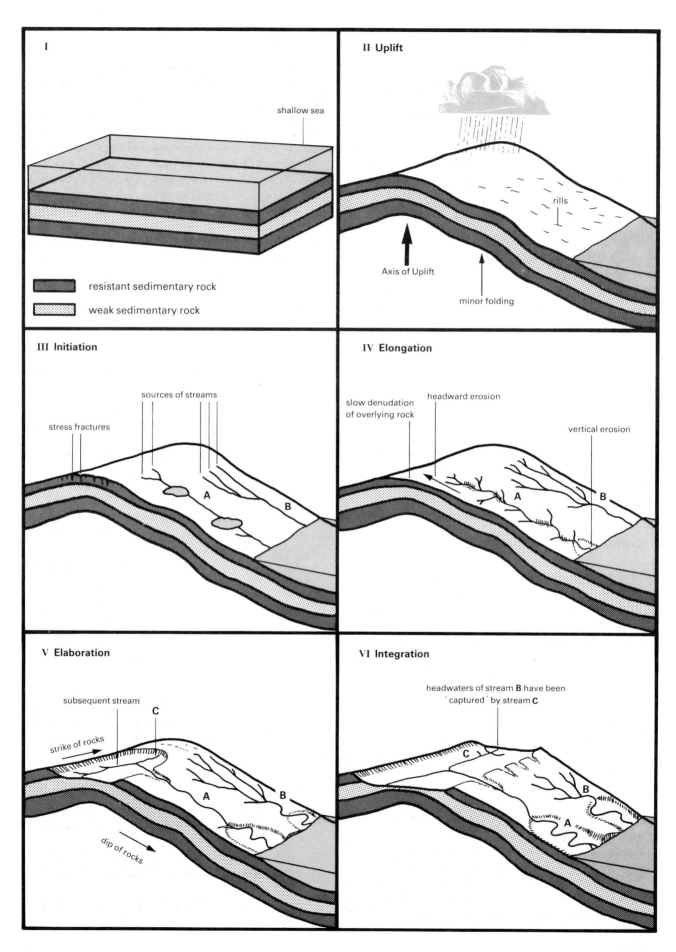

Fig. 5.3 Stages in the development of a river system

The Development of a River System

Before considering in more detail the workings of and relationships existing within river systems, it is necessary to consider how such systems might have developed. The sequence of diagrams Fig. 5.3 I–VI provides a useful basis for study.

Fig. 5.3 I shows an initial situation with alternating resistant and weak sedimentary rocks (for example sandstones and clays) submerged beneath a shallow sea.

Fig. 5.3 II – *uplift* – shows the result of major tectonic movements which have raised a large anticline, with some minor folding, above sea level. Rain falling over the new land mass will, in time, find its way to the sea, either by surface run-off or by through-flow, the slow seepage of water through joints and pores in the rocks (see Fig. 5.12). In the early stages following uplift, the running water on the surface flows to the sea in a disorganised series of rivulets or **rills**. However, in Fig. 5.3 III can be seen the first signs of a river system – *initiation*. Here a major **consequent** or **dip slope stream** (A) has developed, with its course controlled by (i.e. a direct 'consequence' of) the initial slope and dip of the rocks. You will notice that the minor folding has resulted in the development of temporary lakes. Consequent stream B shows another form an early stream system may take, with the tributaries joining at a very acute angle. A final point concerns the fissures and fractures on the crest of the anticline, the result of enormous stresses on the rock during folding; they are being slowly exploited by weathering processes (see Chapter 2).

In Fig. 5.3 IV the streams begin to lengthen – *elongation* – as the sources cut steadily back into the hillside by a process known as **headward erosion**. In addition downcutting or **vertical erosion** has produced a small valley. You should notice that the topmost resistant stratum is slowly being denuded.

Fig. 5.3 V shows the river system becoming much more fully developed – *elaboration*. The headwaters of stream A have managed to erode vertically through the overlying resistant rock to the weaker rock below. These upper tributaries are known as **subsequent** or **strike streams**, having adapted themselves to the different structure of the rock on which they are flowing. Subsequent stream C can be seen to be steadily eroding headwards. This diagram also shows that the valley which appeared in Fig. 5.3 IV has now broadened; this is the result of the stream, now close to sea level, meandering sideways and so performing more **lateral erosion** than vertical erosion.

Fig. 5.3 VI demonstrates the truly dynamic nature of river systems, as the headwaters of consequent stream A have managed to attract and divert the headwaters of stream B. This merger of river systems – *integration* – is undertaken by a process known as **river capture** or **piracy**, details of which are considered later in the chapter.

The Drainage Basin

The drainage basin represents the entire area drained by a particular river system, that is a major stream and all its tributaries. If not lost by evaporation or through underground percolation (see Fig. 5.12), all rain falling within the basin will eventually drain to the sea as 'run-off' along the network of streams. The boundary line dividing one drainage basin from another is termed the **watershed** and this may be identified on a map by drawing a line between the highest points which separate streams flowing into neighbouring basins. Fig. 5.4 shows this; the watershed between river systems A and B follows the crest-line of the ridge and its lower-lying col. You will notice that the sources of the streams do not extend back to the watershed itself, as a certain zone is necessary to enable enough water to be collected to form a stream. Drainage basins often have a characteristic 'pear-shape'.

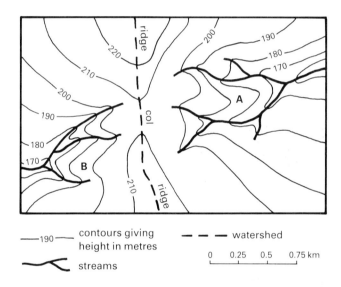

Fig. 5.4 The position of the watershed

Fig. 5.5 shows the watershed of the Upper Wye which rises just to the south-east of Plynlimon in Mid Wales. Neighbouring drainage basins provide the headwaters for the Rivers Severn, Rheidol and Ystwyth.

3 a) Study Fig. 5.5 and notice how the watershed picks its way between the tributaries of the different basins.
b) Locate the region in your atlas and observe the contrasting courses of the different rivers to the sea.
c) Show these courses by means of a sketch map.

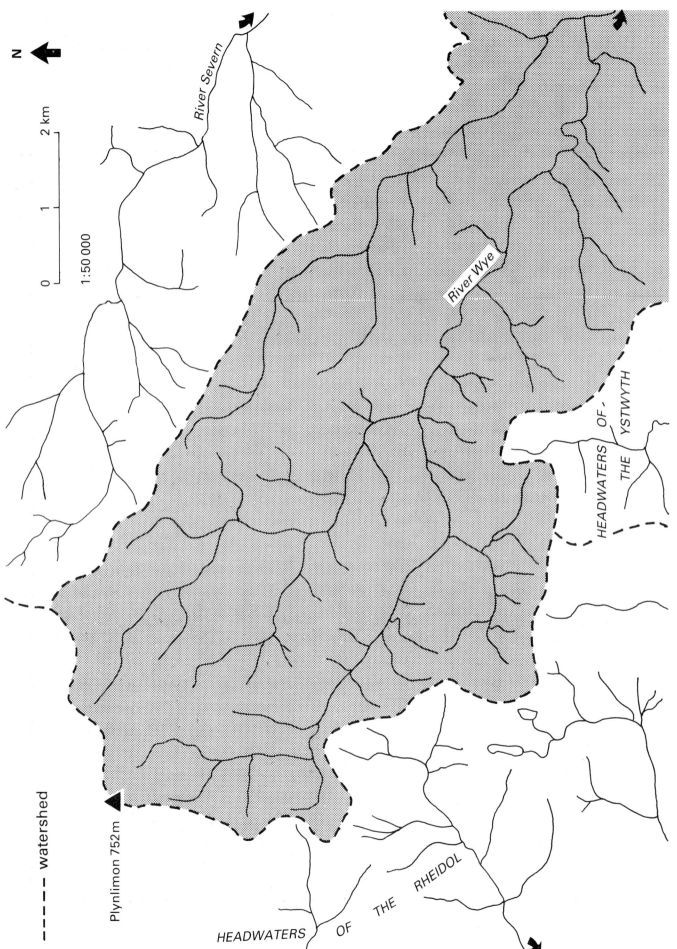

Fig. 5.5 The watershed of the Upper Wye and neighbouring catchments

River Processes and Associated Landforms

Like other disciplines, geomorphological study involves a routine procedure of **observation**, **description**, followed by **explanation**. Having observed the nature of the Upper Wye headwaters it now becomes necessary to describe them. Fig. 5.6 shows four common types of drainage pattern which can be found in the landscape. Such patterns may occur on a great variety of scales, and river systems may exhibit examples of more than one pattern.

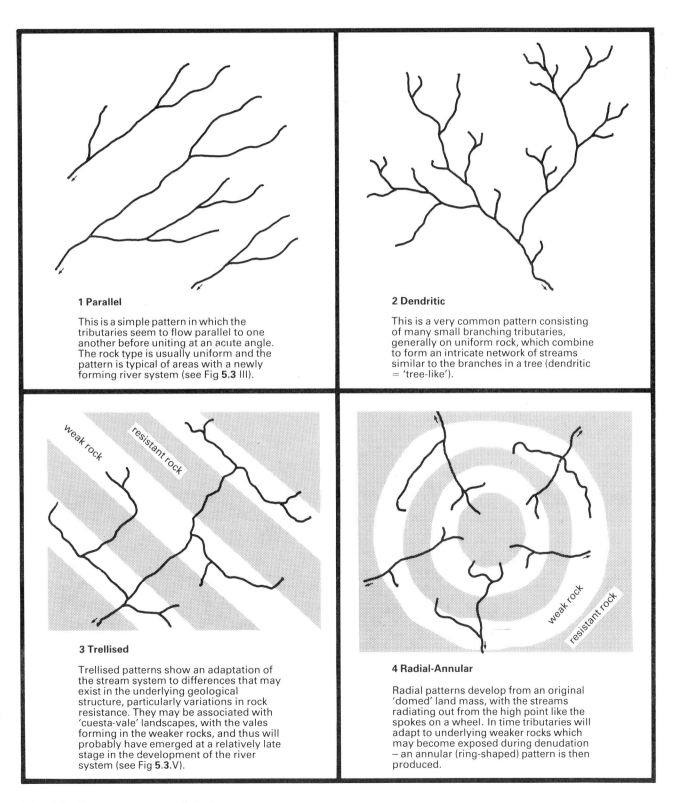

1 Parallel

This is a simple pattern in which the tributaries seem to flow parallel to one another before uniting at an acute angle. The rock type is usually uniform and the pattern is typical of areas with a newly forming river system (see Fig 5.3 III).

2 Dendritic

This is a very common pattern consisting of many small branching tributaries, generally on uniform rock, which combine to form an intricate network of streams similar to the branches in a tree (dendritic = 'tree-like').

3 Trellised

Trellised patterns show an adaptation of the stream system to differences that may exist in the underlying geological structure, particularly variations in rock resistance. They may be associated with 'cuesta-vale' landscapes, with the vales forming in the weaker rocks, and thus will probably have emerged at a relatively late stage in the development of the river system (see Fig 5.3.V).

4 Radial-Annular

Radial patterns develop from an original 'domed' land mass, with the streams radiating out from the high point like the spokes on a wheel. In time tributaries will adapt to underlying weaker rocks which may become exposed during denudation – an annular (ring-shaped) pattern is then produced.

Fig. **5.6** Common types of drainage pattern

4 Study Figs. 5.5 and 5.6 together.
a) Is the pattern of the Upper Wye adequately represented by any one of the four patterns?
b) Study your local 1:50 000 Ordnance Survey map and carefully trace off the streams of your local drainage basin (you may find it necessary to join together two neighbouring sheets in order to do this). You will need this for the questions below and exercises 6 and 8.
c) What type of drainage pattern do you observe?
d) Using the method outlined above, delimit the watershed of your basin, being careful to establish exactly which way the land is sloping on the map.
e) Compare your river system with that of the Upper Wye and describe as accurately as possible the differences (and similarities) between the two systems.

Stream Ordering

A river system consists of a number of **stream segments**, stretches of a stream between stream junctions. Ordering is a way of establishing the relative importance in the river system of one stream segment compared with another.

All stream segments with a source are identified as 'first order' segments. When two first order segments meet, a 'second order' segment is formed; two second order segments meeting create a 'third order' segment ... and so on. It should be noted that first order segments can be tributaries of higher order segments (second, third, etc.) without affecting the order value of those segments. *A higher order stream segment is only created when two stream segments of the same order meet.* The drainage basin itself can be given an order; this is simply the value of the highest order stream segment it contains, e.g. a third order drainage basin.

Since different people will describe these drainage patterns in different ways, greater accuracy and consistency of description can be achieved by actually measuring some of the drainage basins' characteristics. **Stream ordering** or **order designation** is one such measurement technique.

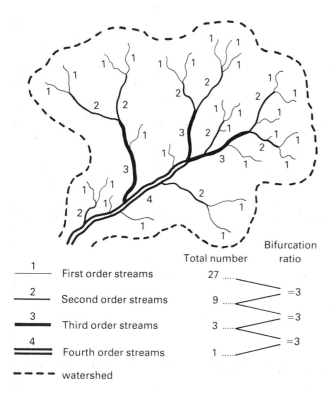

Fig. **5.7** Stream ordering in an idealised drainage basin

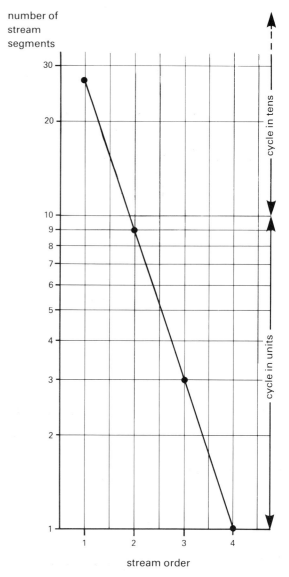

Fig. **5.8** Graph showing the relationship between stream order and number of stream segments in the idealised drainage basin

River Processes and Associated Landforms

Fig. 5.7 shows the result of stream ordering in an idealised drainage basin using the procedure outlined above. This is a fourth order basin.

Higher order streams are far less frequent in number than lower order streams. The ratio of the number of streams of one order compared with the number of streams of the next highest order is known as the **bifurcation ratio**, calculated simply by dividing one by the other. The idealised example in Fig. 5.7 shows a constant bifurcation ratio of 3 between all the orders.

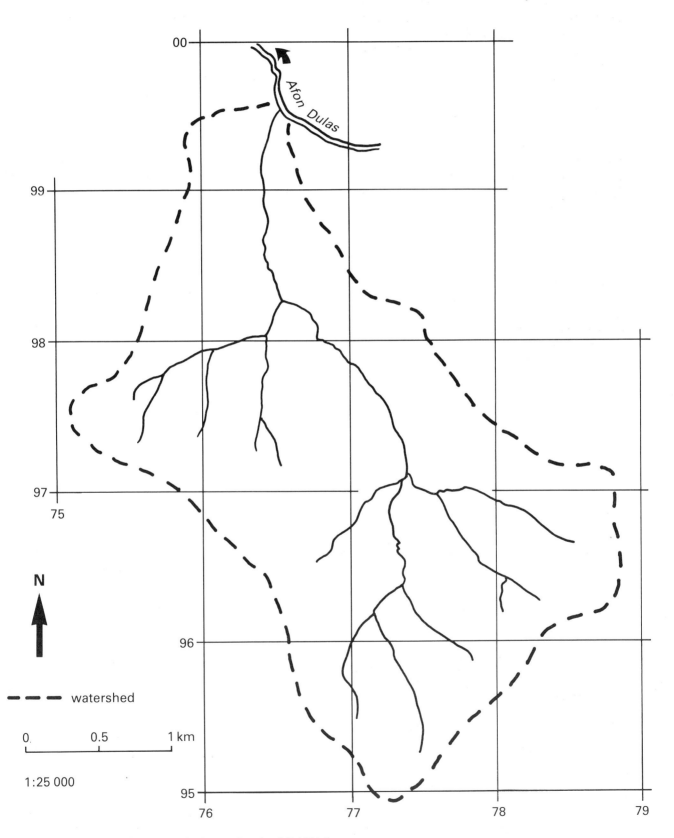

Fig. **5.9** The Nant Ceniarth drainage basin, Mid Wales

Such ratios tend to be remarkably constant within actual drainage basins too, though variations do exist *between* basins.

Semi-logarithmic graph paper, with a normal (arithmetic) horizontal scale and a logarithmic vertical scale arranged in cycles, can be used to show these relationships. Each vertical cycle consists of lines, progressively closer together, numbered 1 to 9. They can be used to represent units, tens, hundreds, etc. as desired; successive cycles will thus begin 1, 10, 100 etc. This is shown in Fig. 5.8. When plotting stream numbers against order, the logarithmic scale is particularly useful and effective. It enables the constant bifurcation ratio, featured in Fig. 5.7, to be represented as a perfectly straight line, rather than a steeply rising curve that would have been produced using an arithmetic scale. It therefore shows very clearly that river systems develop in a *regular* rather than random way.

> **5** Study Fig. 5.9 which shows the drainage basin of the Nant Ceniarth, one of the headwater streams of the River Dovey in Wales.
> a) On a copy of the map carry out the ordering of the streams.
> b) What order basin is this?
> c) Count up the number of streams of each order and plot your results on semi-logarithmic paper, as shown in Fig. 5.8. How straight is the line produced?
> d) Calculate the bifurcation ratios.

The Nant Ceniarth is not a perfectly balanced basin, but since stream ordering provides a standardised way of examining basins, all members of your class *should* produce the same results. An important feature of a technique such as this is that, by plotting results from two drainage basins on the same graph, any differences between them will be shown up clearly.

> **6** a) Repeat the same procedure of ordering and graphical representation for your own basin, which you traced off in exercise 4(d).
> b) What differences do you observe between your basin and that of the Nant Ceniarth?

In addition to counting numbers of stream segments, other map measurements are possible which will further emphasise the organisation of streams in a drainage basin. For example, the relationship between the mean (average) length of stream segments and their order has been observed to follow a regular progression. The higher the stream order, the greater the mean length. This is shown in Fig. 5.10.

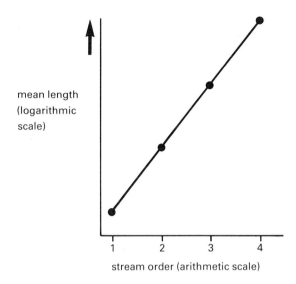

Fig. **5.10** Graph showing the relationship between stream order and mean length of stream segments in a drainage basin

Stream length can be calculated from a map by running a strip of paper (or piece of string) alongside the stream segment concerned, carefully taking into account all the twists and turns. The measured length can be read off against the map scale.

> **7** a) Return to Fig. 5.9. Give each of the stream segments a code by order, e.g. 1a, 1b, 2a, 2b, etc. Measure the length of each segment and prepare a table of results. Calculate the mean length for each stream order.
> b) Plot your results on semi-logarithmic graph paper. Comment on the regularity of the results. Does any order have a mean length different from what you would expect?

Your main conclusion from all this work is that river systems, however irregular and haphazard they may appear, do show a strong degree of organisation and balance, particularly in terms of a) the number of streams they contain, and b) the length of those streams.

Drainage Density

Another measure of a drainage basin that is commonly used is **drainage density**, which expresses the total

River Processes and Associated Landforms

Basin no.	Basin name	I Area (km²)	II Stream length (km)	III Drainage density (km/km²)	IV Permeable rock?	V Annual rainfall (mm)
1	Ash Bourne	18.7	37.6	2.01	no	859
2	Darwell	9.6	17.2	1.79	no	904
3	Nunningham Stream	16.9	28.0	1.66	no	807
4	East Dart	21.5	31.5	1.47	no	2103
5	Ray	18.6	15.0	0.81	no	660
6	Croasdale Brook	10.4	22.5	2.16	no	1839
7	Derwent (Northumb.)	118.0	218.5	1.85	no	932
8	St. John's Beck	40.9	90.8	2.22	no	2710
9	Glaslyn	68.6	194.8	2.84	no	3279
10	Wye	27.2	55.5	2.17	no	2461
11	Babingly	39.6	11.1	0.28	yes	696
12	Foston Beck	57.2	26.5	0.46	yes	719
13	Great Eau	77.4	52.0	0.67	yes	635
14	Wallop Brook	53.6	6.9	0.13	yes	803
15	Axe	12.0	9.0	0.75	yes	1117

Table **5.1** Measurements of selected drainage basins in Britain

length of streams draining a particular unit area of the basin. It is calculated using the following simple formula:

$$\text{Drainage density} = \frac{\text{total length of streams in km}}{\text{area of basin in km}^2}$$

The measurement of stream length has been described above. Calculating the area of a drainage basin may prove awkward since its shape may be quite irregular. Follow this procedure:
a) Note the scale of the map, usually 1:50 000 or 1:25 000, and prepare a tracing of the basin, marking the watershed clearly.
b) Superimpose this tracing on metric graph paper and count up the total number of small (i.e. 2 mm) squares that fall within the watershed. When the watershed cuts across the small squares some intelligent guesswork will be needed to estimate the total the parts add up to. At a scale of 1:50 000, one 2 mm square represents 1/100 of a square kilometre whilst at a scale of 1:25 000 one 2 mm square is 1/400 of a square kilometre.

8 a) Refer back to Fig. 5.9 and the data you collected in the previous exercise on stream lengths in the Nant Ceniarth basin. Using the method described above, calculate i) the area of the basin and ii) its drainage density. Notice that grid lines have been marked to assist you.
b) Calculate the drainage density for your own drainage basin, traced off in exercise 4.

Drainage density is a particularly good measure which allows comparisons between drainage basins to be made.

9 Study the drainage density figures in Table 5.1 (column III), given for a number of selected drainage basins in Britain.
a) Using the figures for basin area and stream length shown in columns I and II, plot each basin on a scattergraph. A start has been made for you in Fig. 5.11. Be careful to number each point as in the table, so that the basins can be readily identified on the graph.
b) Having completed the plots, draw on a straight line which you consider best reflects the relationship between stream length and basin area; in this case it will slope upwards from left to right. Try to position the line so that an equal number of points lie on each side. This line is known as the **best-fit** or **regression line** and depicts the average stream length : basin area relationship (i.e. drainage density). Fig. 3.15 on page 51 will help you establish the strength of this relationship.
c) On the graph mark on the results you obtained for the Nant Ceniarth basin and your own basin.

By studying the pattern of points on the scattergraph it is clear that they do not all lie on the best-fit line, which would only happen if all the basins had the same drainage density. Consider, for example, points 9

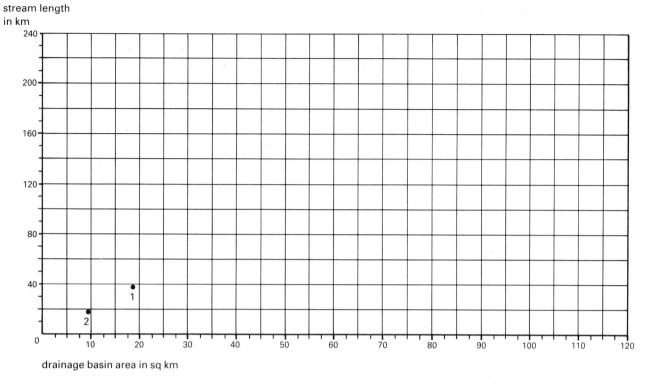

Fig. **5.11** Scattergraph showing the relationship between stream length and basin area for selected basins in Britain

(Glaslyn) and 13 (Great Eau); though their basin areas are approximately the same, the stream length of Glaslyn (and therefore its drainage density) is much greater than that of Great Eau. A feature of this particular measure of a river system, like that of stream ordering, is that it highlights differences which may exist between drainage basins. Geomorphologists need to explain why such differences occur and this means looking for possible variations in, for example, rock characteristics and annual rainfall.

An important characteristic of a rock is its **permeability**, the capacity to allow water to seep through via joints and pores (studied in Chapter 2). As rainfall is the major source of stream water, then if a rock were exceptionally permeable it is reasonable to presume that no streams would be flowing on the surface (assuming nothing intervened to prevent water seepage). Furthermore if a basin were to receive no rainfall then we would expect no streams to exist, no matter what the permeability of the rock. From these

Basin no.	Basin name	Annual rainfall (mm)	Rank	Drainage density (km/km^2)	Rank	d	d^2
1	Ash Bourne	859	8	2.01	5	3	9
2	Darwell	904	7	1.79	7	0	0
3	Nunningham Stream	807		1.66			
4	East Dart	2103		1.47			
5	Ray	660		0.81			
6	Croasdale Brook	1839		2.16			
7	Derwent (Northumb.)	932		1.85			
8	St. John's Beck	2710		2.22			
9	Glaslyn	3279		2.84			
10	Wye	2461		2.17			

Table **5.2** Comparing the influence of annual rainfall on drainage density for selected drainage basins (see exercise 11)

River Processes and Associated Landforms

lines of thought we can construct two simple hypotheses:

I – *that those basins with permeable rock will have lower drainage densities than those with impermeable rock.*

II – *that the greater the annual rainfall total of a basin the greater will be the drainage density.*

10 Test hypothesis I:
a) Refer to Table 5.1, column IV and on your scattergraph draw a ring around those basins with permeable bedrock (i.e. basins 11–15).
b) Remember that the best-fit line picks out the average relationship between area and length, so points above the line will have higher than average stream length for their area (a greater drainage density) whilst points below will have a lower than average stream length.
c) What is particularly noticeable about the basins you have ringed in relation to the best-fit line?
d) Does their position help you to accept or reject the hypothesis?
e) From the positions of the points for the Nant Ceniarth basin and your own basin, can you suggest the likely permeability of the rocks?

11 a) Find the rank position for each basin in relation to the rest, for both rainfall and drainage density. Thus Ash Bourne with the eighth highest rainfall total of the ten basins is ranked 8. Enter your rankings on a copy of Table 5.2.
b) Complete the d and d^2 columns (d is the difference between the two rankings for each basin, whilst d^2 is that difference squared).
c) Add up the d^2 column to find Σd^2 (Σ is 'the sum of').

Substitute Σd^2 in the following equation:

$$r = 1 - \frac{6 \Sigma d^2}{n^3 - n}$$

where r = the correlation coefficient,
n = the number in the sample (i.e. 10 basins).

What value of r have you obtained?
d) Interpret the result. If calculated correctly your correlation coefficient will lie between -1 and $+1$. The closer to ± 1, the stronger the relationship between the two variables, whilst the closer to 0, the weaker the relationship. A negative answer indicates an **inverse relationship**, where an *increase* in the independent variable produces a *decrease* in the dependent variable (not so in this case). How strong is the relationship between rainfall and drainage density?
e) Interpret this *in terms of your hypothesis*.

We will now test hypothesis II. As we have shown that rock permeability can affect drainage density, it would seem a sound idea when testing for the link between rainfall and density to ignore those basins with permeable rock, as they may distort the result (i.e. concentrate only on basins 1–10).

A scattergraph can also be used in this exercise, with rainfall on the horizontal axis and drainage density on the vertical axis (see Table 5.1, columns III and V). However, another method of measuring the strength of the relationship between these two variables is to calculate the **correlation coefficient**. As our hypothesis assumes that rainfall influences density, annual rainfall may be regarded as the **independent variable**, with drainage density as the **dependent variable**, requiring, i.e. 'depending' on, a change in rainfall total before being able to change itself.

One of the simplest ways of calculating the correlation coefficient between two variables is to use the Spearman's Rank method. This can be done for annual rainfall and drainage density for the ten sample streams: follow the procedure outlined below in conjunction with Table 5.2.

The Work of Individual Streams

A consideration of **stream discharge** is of fundamental importance when examining the processes at work in a stream. Stream discharge is the amount of water actually flowing in a stream past a given point and is usually measured in litres per second or cubic metres per second (1000 l = 1 m^3).

Fig. 5.12 shows where the water in streams comes from:

a) groundwater storage within the bedrock – this produces **base-flow** which remains fairly constant throughout the year, with some seasonal variations.
b) surface run-off, also known as **overland flow**; this will vary with the level of rainfall at any time.
c) infiltration through the soil and sub-soil, known as **through-flow**; this is related to rainfall too.

A glance at any river system (e.g. Fig. 5.5) reveals that as streams tend to combine, stream discharge will increase downstream.

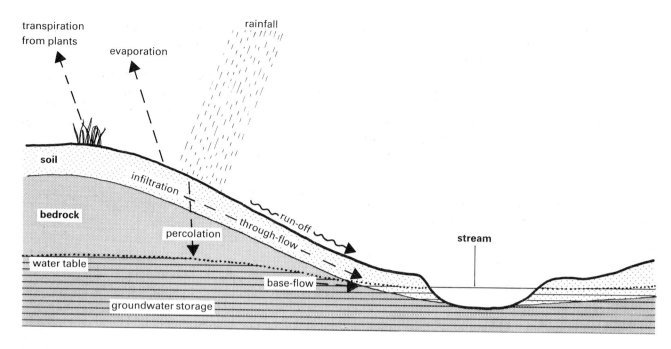

Fig. 5.12 Contributions to stream discharge

Measuring Stream Discharge

The main problem in measuring stream discharge is the need to take account of variations in discharge that exist within the stream channel. Fig. 5.13 shows that discharge decreases a) with depth, and b) towards the stream banks; this is due to friction where the stream water meets the channel bed and banks.

You will need the following equipment to measure discharge:

a) a tape measure
b) a stop watch
c) an orange (to act as a float)
d) two surveyor's measuring poles.

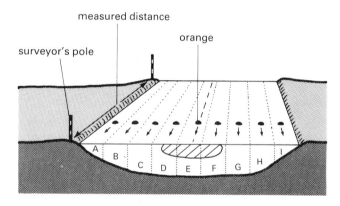

zone of highest velocity and greatest discharge per unit area

Fig. 5.14 Measuring stream discharge

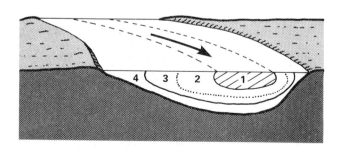

1 – Zone of highest velocity and greatest discharge per unit area
2 – Above average velocity and discharge per unit area
3 – Below average velocity and discharge per unit area
4 – Zone of lowest velocity and least discharge per unit area

⋯⋯ line showing average discharge per unit area

Fig. 5.13 Variations of discharge within the river channel

It is first necessary to select a suitable stretch of stream, preferably straight, and to establish the cross-profile and the cross-sectional area of the stream channel. Working across the stream and using a surveyor's pole, the depth of water is recorded at regular intervals (e.g. every metre); the pole should not be pushed into any mud or gravel.

Secondly, a certain length of the bank has to be measured (say 10–20 metres). With the width of the stream broken up into conveniently sized sections, an orange is floated down the centre of each section and timed accurately in seconds over the measured bank distance. Fig. 5.14 illustrates this, with nine sections (A–I) being chosen, though a smaller number is adequate. By repeating timings over the same section an average timing can be calculated. From this the speed (velocity) of stream flow in metres per second can be established using the simple formula:

$$\text{velocity} = \frac{\text{distance in metres}}{\text{time in seconds}}$$

Back in the classroom, it is necessary to draw the channel profile on graph paper from the data collected, selecting a suitable scale. By counting the number of squares occupied by the profile on the graph paper and converting this to square metres, the cross-sectional area for the channel and each of the sections can be found. Discharge is then found using the formula: discharge = cross-sectional area × velocity. However, because we saw above that discharge varies with depth, it is customary to multiply the result by 0.8 to take into account the frictional drag. The formula then becomes:

discharge = cross-sectional area × velocity × 0.8

Adding up the discharge figures for each section gives the total discharge of the stream.

It is worth examining what effect changes in rainfall will have on a) discharge, and b) the work undertaken by streams. Fig. 5.15 attempts to show this for the River Otter in Devon. The diagrams show the measured effects of two storms, 1 and 2.

> **12** Study Fig. 5.15 A and B and answer the following questions:
> a) Exactly how much rain fell during each of the storms?
> b) How long did each storm last?
> c) What was the discharge before storm 1 and what were the probable origins of this water? (See Fig. 5.12.)
> d) What was the peak discharge following storm 1?
> e) What was the percentage increase in discharge following storm 1?
> f) The peaks of the discharge curve do not coincide with the peaks of rainfall. What is the approximate lag-time between the two?
> g) Refer back to Fig. 5.12 and suggest a few reasons why a lag-time should exist.

Fig. 5.15 A and B combined is known as a **storm hydrograph**.

Your answers to these questions should illustrate key features of stream behaviour following rainstorms. The effects on the work that streams can undertake are just as pronounced.

> **13** Study Fig. 5.15 A and C. In C the amount of sediment being transported by the stream is measured as a percentage of the volume of water. Answer the following:
> a) Do the peaks for discharge and amount of sediment load coincide?
> b) What is the percentage increase in sediment load as a result of storm 1? (Note that the load carried following storm 2 is reduced because most of the material would already have been carried away by storm 1.)

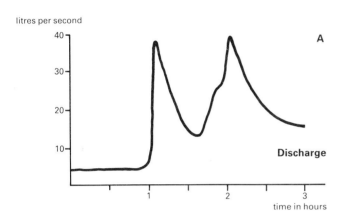
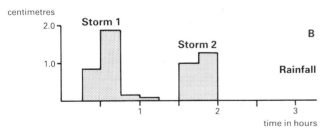
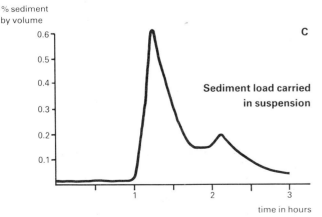

Fig. **5.15** The effect of two storms on discharge and sediment load of the River Otter, Devon

In times of flood, therefore, streams are able to transport much more material than they can under normal circumstances. Thus fluctuations in discharge have important implications for the amount and type of work that a stream can do.

Load, Erosion and Deposition

In August 1952 a severe storm over Exmoor (23 cm of rain!) caused the normally small East and West Lyn rivers to flood at Lynmouth. The peak discharge was 511 m³/sec; in Britain only the River Thames has ever exceeded this figure.

> **14** Locate the Lyn rivers and the Thames in your atlas. Compare the sizes of the two drainage basins. This should give you some idea of the severity of the storm.

The enormous increase in volume and velocity gave the rivers a great deal of extra energy, causing tremendous damage. The West Lyn, for example, was estimated to have moved 50 000 tonnes of boulders in 24 hours, with one 7.5 tonne boulder actually being found in a hotel basement! Fig. 5.16 shows some of the after-effects of the flood.

Fig. 5.17 is a graph illustrating the relationship between particle size and stream velocity and is of value in determining what processes are actually occurring within the stream channel. You should note the following points:

a) For curve 1, the general rule is that the larger the particle, the greater the stream velocity required to erode it, *but* below about 0.6 mm particles become so small that they are not easily eroded – they tend to stick to one another and to the channel bed.

b) For curve 2, as velocity falls, so particles are deposited in decreasing order of size.

> **15** Study Fig. 5.16 once again and refer to Fig. 5.17.
> **a)** Estimate the diameter of the large boulders using the men as a rough scale.
> **b)** Estimate the velocity of the Lyn that was required to shift them. (You will have to extend curves 1 and 2 on Fig. 5.17; notice that the scales are increasing geometrically.)
> **c)** Explain why so many boulders were deposited in the road shown in the photograph. (Refer back to Fig. 5.15 and look at the sudden changes in the steepness of curves A and C.)

Fig. 5.16 The effects of flooding on Lynmouth

River Processes and Associated Landforms   113

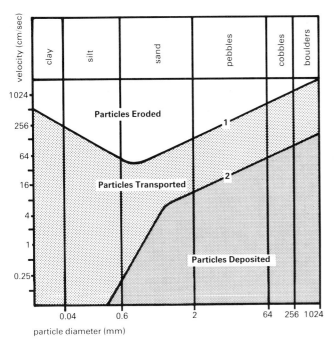

Fig. **5.17** The relationship between stream velocity, particle size and process

One of the influences on stream velocity is gradient; as gradient increases so will velocity, other things being equal. Fig. 5.18 shows the gradient curve for the River Clarach which flows into Clarach Bay on map extract 2; this curve is known as the **long profile**. It has been drawn from the map by running a strip of paper carefully along the river from source to mouth, including all twists and turns, and noting on the paper precisely where contour lines cross the river and what their value is. (A magnifying glass is often very helpful for this exercise.) From the data on the strip of paper the long profile can be drawn, in the same way as cross-profiles are drawn. It is customary to show the profile sloping from left to right, even though on the map the River Clarach is flowing from east to west. The vertical scale of the profile may be drawn slightly larger than the horizontal scale for diagrammatic purposes.

16 a) Using Fig. 5.18 describe the shape of the long profile, noting particularly the change in gradient of the River Clarach from its source to its mouth.
b) From map extract 2, draw the long profile of the tributary of the River Clarach which has its source at 662872. For your profile you should use a vertical scale of 2 mm for 15.24 m (the contour interval). The horizontal scale will be the same as for the map.
c) Compare the shapes of the two profiles. What similarities do they reveal?

At four points, identified A, B, C and D on Fig. 5.18, the mean velocity was determined. The results are shown on the graph in Fig. 5.21. Far from showing a drop in velocity as might be expected, the graph actually shows an increase, slackening only slightly towards the sea. The reason for this centres around the character of the river channel. In the upper course the channel cross-profile is very irregular and so a lot of stream energy is lost in friction with the bedrock, thus restricting velocity. In the lower course the channel is smoother, energy loss is therefore less and velocity remains fairly constant.

If you relate the mean velocities of the River Clarach to Fig. 5.17 you will note that the most common process is the transportation of particles. This is quite characteristic, since geomorphologists have long regarded the shape of the long profile as having evolved for just this purpose. A smooth, concave long profile is described as being **graded**. We should not forget however, that Fig. 5.21 refers to *mean* velocities; *actual* velocities at a given time are likely to differ from the figures quoted.

We need to consider the processes of transportation, erosion and deposition in greater detail, with particular reference to stream load. A stream's load can be derived from a) outside the channel, as a result of weathering and gravity movements in the river valley (see Chapter 2), or b) within the channel. Because there will be many different-sized particles in the channel, the way they will be transported downstream will vary. When particles are small in relation to stream energy, they will be carried in **suspension** within the stream water itself. Conversely, when particles are large, they will be dragged along the channel bottom forming the **traction** load. When particle size is marginal for either suspension or traction (i.e. medium size), that particle will be bounced along the stream bed as part of the **saltation** load. Finally, because some rock minerals dissolve in stream water, mention must be made of the **solution** load.

It has been estimated that the total annual load carried to sea by the Mississippi River in the USA amounts to nearly 540 million tonnes, of which 63% is carried in suspension, 29% in solution, with the remaining 8% by traction and saltation.

When stream energy is high in relation to its load, erosion will occur. Flowing water itself will lever up slabs and loosen particles of rock on the channel bed and banks; this is known as **hydraulic action**. However, most of the erosion is done by the material being transported in the stream. Particles will continually be hitting the bed and banks causing them to erode gradually; this is called **corrasion**.

Pot-holing is a particular form of corrasion. A pebble will become lodged in an irregularity on the stream bed and get swirled round by the water. The swirling action acts like a drill, producing a hole which may be enlarged by amalgamation with adjacent holes. Fig. 5.20 shows a

height in metres
above sea level

Note: Vertical exaggeration = 6.56

Fig. **5.18** Long profile of the River Clarach, Mid Wales

height in metres
above sea level

Note: Vertical exaggeration = 6.56

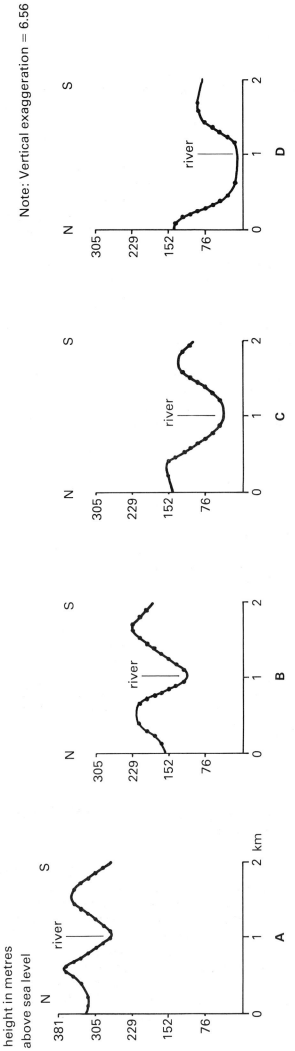

Fig. **5.19** Cross-profiles of the valley of the River Clarach

River Processes and Associated Landforms

sequence of photographs taken on the River Rheidol in its upper course when the discharge was quite low. In A pot holes are visible and their size can be estimated by comparison with the sitting figure. In B, taken 100 metres downstream, pebbles can actually be seen on the right of the picture, still trapped in the exposed pot holes; whilst in C, 50 metres further downstream, the amalgamation of pot holes has produced a small but narrow gorge. A further point emerging particularly from B is the great channel irregularity, mentioned above, which slows down stream velocity. Pebbles trapped in pot holes, like those being transported along the stream bed, will eventually be worn away themselves; this erosion of the load is known as **attrition**. The erosion process which involves the dissolving of bedrock and the removal of the load in solution is known as **corrosion**.

An important influence on stream erosion is gravity, which pulls water downslope to the sea. The pull of gravity is strongest in the upper course, as the stream here is higher above sea level; as a result erosion is mainly vertical. Near the mouth there will be little or no energy resulting from height, as the stream is close to sea level; vertical erosion is impossible so, given the right conditions for erosion (e.g. flooding), sideways or lateral erosion takes place (refer back to Fig. 5.3). These differences in erosion along the stream course have important consequences for landform development.

Finally, when stream energy is low in relation to its load, deposition will occur. Like erosion, deposition can take place at any point in the stream's course. In the upper course, pebbles and boulders may litter the channel, whilst in the lower course the main effects of deposition are found on the broad flat floor of the **flood plain**. This is covered in fine grained sediment termed **alluvium** which has been deposited following an overflow of the stream during flooding; beyond the banks, stream energy is spread over a large area instead of being concentrated in the main channel, and so deposition occurs.

Landforms Associated with River Processes

The Valley

The relationship of river processes to valley form can be appreciated by referring to the sequence of diagrams in Fig. 5.19. This is a series of cross-profiles of the valley of the River Clarach, drawn at the points where the mean velocity readings were determined; they have been drawn to the same vertical exaggeration as the long profile (Fig. 5.18). The features these profiles reveal are in many ways typical of river valleys in general, though the resistant nature of the bedrock in the region has certainly been of some importance.

Fig. **5.20** Pot holes in the Upper Rheidol

Valley cross-profile Grid reference	A off extract	B off extract	C 637838	D 608838
Altitude (m)				
Depth (m)				
Width (m)				
Gradient of north side				

Table **5.3** Cross-profile features of the valley of the River Clarach

17 Study Fig. 5.19 A–D and for each profile carry out the following measurements as accurately as you can. Use Fig. 5.22 as a guide and complete a copy of Table 5.3.
a) How high above sea level is the river?
b) What is the approximate depth of the valley?
c) What is the approximate width of the valley just above river level?
d) Calculate the average gradient for the north side of the valley. The gradient between two points is the ratio of the vertical interval (V.I.) to the horizontal equivalent (H.E.), i.e.

$$\text{gradient} = \frac{\text{V.I.}}{\text{H.E.}}$$

Vertical interval is the difference in altitude between the two points; horizontal equivalent is the horizontal distance between the two points.

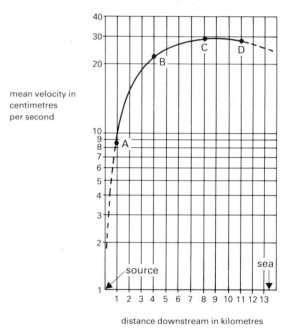

Fig. **5.21** Graph showing mean velocity of the River Clarach along its course

The main value of data collected in this way is that it allows comparisons to be made between valley forms at different locations.

18 a) What has happened to i) valley depth, and ii) valley width, between A and D? Relate these changes to height above sea level.
b) How has the ratio of valley width to depth changed downstream?
c) Have the changes in the gradient of the valley side been significant?
d) What has happened to the cross-sectional area of the valley? (For a rough guide, use width × depth.)

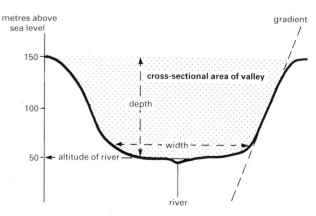

Fig. **5.22** Measurements of the valley cross-profile

River Processes and Associated Landforms

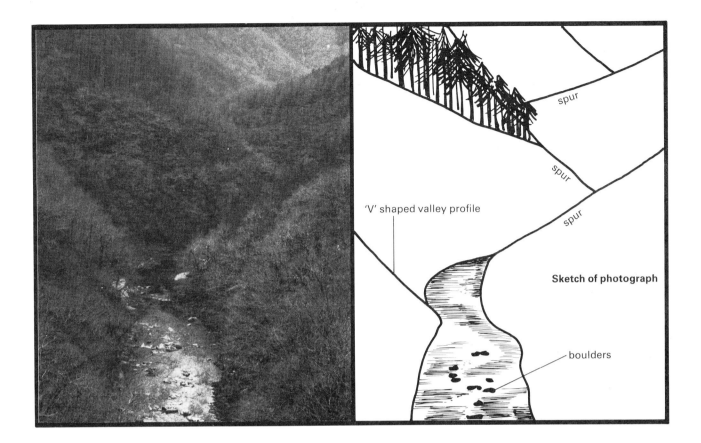

We can now examine more closely some of the distinctive landforms to be found *within* the river valley.

Interlocking spurs and Waterfalls

We have established that there is more vertical than lateral erosion in the river's upper course. This produces a characteristically 'V' shaped valley, well incised into the landscape. However, it must be remembered that not all the cross-sectional area of the valley (in Fig. 5.19 A for example) would have been removed by stream erosion alone. **Sub-aerial processes** of weathering and mass movements (downslope gravity movements) would have played a significant part in removing material from the valley side (see Chapter 2), thus producing a load which is used in the erosion process within the stream. The more active sub-aerial processes are, compared to vertical erosion, the gentler will be the valley side, and vice versa; so here the natural resistance of the rock is an important factor in helping to determine how quickly these sub-aerial processes will operate – the harder the rock, the slower it will tend to break down.

Even in sections close to a river's source, it is likely that some lateral movement will take place; straight stretches of stream are rare. This movement produces ridges or 'spurs' of land projecting out from either side of the valley. They appear to interlock with one another, hence the term **interlocking spurs**.

Fig. 5.23 highlights a number of valley features. The 'V' shape is clearly in evidence, as are the interlocking

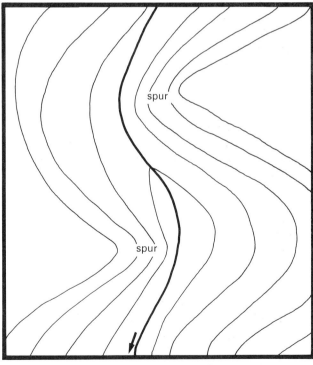

Fig. **5.23** The valley of the River Rheidol

spurs for which a sketch diagram and idealised contour diagram have been included. Notice also the boulders in the stream; how would you explain their presence?

Waterfalls occur either as a result of special geological conditions, or as a result of disturbances in the river's long profile (such as those following glaciation or changes in sea level).

Fig. 5.24 shows the three common types of geological condition which lead to waterfall development. In A horizontally-bedded rocks are overlain by a resistant **cap rock**. When this is breached the weaker rocks underneath are more easily eroded by the falling water and its load, leading to undercutting and the development of a **plunge pool**. As the falls recede, through the process of headward erosion, a gorge is left.

The Victoria Falls (Fig. 5.2) provide a classic example of this type of waterfall. The Zambezi at this point has an *average* annual discharge of 935 m³/sec (nearly twice that of the peak discharge of the East and West Lyn rivers during the 1952 flood (see p. 112). It has managed to breach a thick resistant cap rock of basalt (a hard volcanic rock) to produce the 108 m high falls. In the last 10 000 years they have only receded 2.4 km despite this high discharge, leaving a distinct gorge evident in the foreground of the photograph.

In Fig. 5.24 B the resistant rock is near vertical. Such rock may be igneous, forming a dyke, or it may be old sedimentary rock contorted by earth movements. The rate of retreat of the waterfall in B is slower than in A, where undercutting of the softer rock can cause the collapse of the cap rock.

In Fig. 5.24 C a gently inclined series of alternating weak and resistant rocks produces rapids or 'cataracts'. Plunge pools are much smaller because the vertical force of the water is not as great.

The three diagrams are representative, but Fig. 5.25 shows three photographs of actual waterfalls.

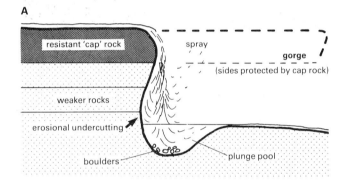

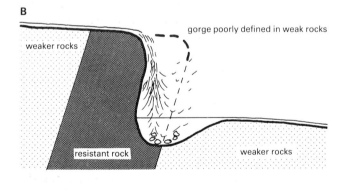

> **19** Study Fig. 5.25 and answer the following:
> **a)** Which of the waterfalls in Fig. 5.24 best describes photograph A? Justify your answer.
> **b)** What is the feature at X?
> **c)** Why is the rock at Y and Z so smooth?
> **d)** Study photograph B where the bedding of the rock is clearly visible. How has this bedding appeared to affect the development of the waterfall as the stream tries to downcut?
> **e)** Study photograph C. What type of waterfall is represented here? Identify the key evidence on the photograph for your conclusion.
> **f)** What other erosional features can be seen which reflect the downcutting that is taking place?
> **g)** Return to Fig. 1.28 (e) on p. 20 and identify the types of waterfall shown on the geological diagram.

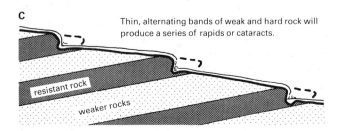

Fig. **5.24** Types of waterfall

Waterfalls may develop in other circumstances, most notably as a result of rejuvenation (page 125) as the stream adjusts to a new (lower) sea level and as a result of glaciation where hanging valleys (page 143) are produced.

A The Mynach Falls at Devil's Bridge

B The Mynach Falls at Devil's Bridge

C Waterfall at Parson's Bridge on the River Rheidol

Fig. **5.25** Waterfalls

Meanders, Oxbow lakes and Deltas

The most characteristic feature of the lower course of streams is the broad flat flood plain shown in Fig. 5.26. Whilst a good deal of deposition occurs here following flooding, the feature owes its origins to lateral erosion by the stream which is close to sea level at this point. Fig. 5.27 shows that the flood plain has been cut into the bedrock and overlain by fine silt and larger (coarser) river gravel deposits.

The development of meanders leads to lateral erosion. Exactly how meanders develop in the first instance is still not entirely understood, but the main current (i.e. that with the greatest velocity) flowing round a meander will gradually be forced to the outside bend, just beyond the crest of the meander. This will cause erosion, particularly in the form of undercutting and slumping, eventually leading to the development of river cliffs, as shown on Fig. 5.27 and at the points marked X on Fig. 5.28. Over time, as Fig. 5.28 shows, the meanders tend to migrate downstream, successively occupying positions 2, 3 and 4 (the same as 1); they become larger as they approach the sea. The effect of this is to produce a whole series of river cliffs, which in time will merge to form distinctive **bluff lines** on the edges of the flood plain.

The undercutting process on the outside bend of the meander results from the particular nature of the water flow. As well as the main current being flung towards the

Fig. 5.26 Flood plain and meanders of the River Cuckmere, Sussex

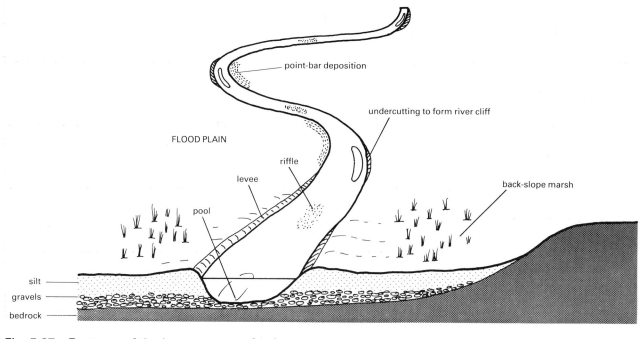

Fig. 5.27 Features of the lower course of a river

River Processes and Associated Landforms

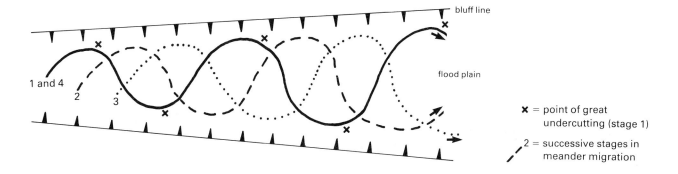

Fig. 5.28 Meander migration and flood plain development

outside of the meander, water tends to flow in a corkscrew-like way. Fig. 5.29 demonstrates this and the effect it has on the erosion of the bank and the development of a relatively deep **pool** within the river channel. Another effect is the build-up of deposits on the inside bend of the meander where the current slackens, resulting in the formation of a **point-bar**. A whole series of such deposits will be left as the meander migrates downstream. Slack currents near the river banks will also lead to deposition of silt, which may accumulate into small embankments or **levees**, particularly following a period of flooding. The effect may be to allow the stream level to exist slightly above the level of the flood plain; seepage of water into the channel is therefore impeded, the flood plain becomes poorly drained or waterlogged and a **back-slope marsh** (or backswamp) is produced. These features are shown on Fig. 5.27.

Measurements	Meander 1	Meander 2
1 Channel width in metres	a = b = c = d =	a = b = c = d =
2 Mean channel width in metres		
3 Meander wavelength (W_1 to W_2) in metres		
4 Ratio of mean width to wavelength		
5 Meander amplitude in metres		
6 Banklength for meander wavelength in metres		
7 Banklength for 1000 metres wavelength		

Table 5.4 Comparing features of two meanders on the River Thames (see exercise 20)

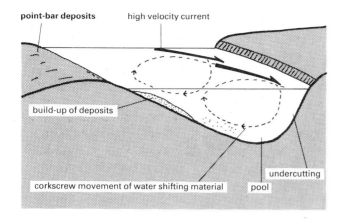

Fig. 5.29 Processes within a meander

As river discharge increases downstream, so changes can be expected in the size and shape of the meanders. A greater discharge means a bigger, wider channel and therefore an increased potential for lateral erosion, leading to a wider flood plain. Certainly, geomorphologists have noticed the regularity of the spacing not only of the pools within the channel, but also of the zones of shallower water known as **riffles** (see Fig. 5.27). This apparent regularity has led to much measurement of meander features.

Fig. 5.30 is the basis for an exercise comparing two meanders in the River Thames, approximately 17 km apart, one upstream from London, the other downstream.

> **20** Study Fig. 5.30 and Table 5.4 carefully. Locate the meanders in your atlas.
> a) Using the measurements guide in Fig. 5.30 A, complete a copy of Table 5.4.
> b) Compare the two meanders. Is there a downstream increase in:
>   i) channel width
>   ii) meander wavelength
>   iii) meander amplitude?
> c) Channel width : wavelength ratios of meanders in other streams are most commonly between 1:10 and 1:14. Do your results confirm this? What might have happened to the banks of the Thames to affect the ratios you have calculated?
> d) From measurements 6 and 7 we can obtain a rough measure of how curved the meander is (i.e. its sinuosity). The greater the banklength for a given wavelength measure, the greater the sinuosity. Has the meandering of the Thames become more sinuous downstream?

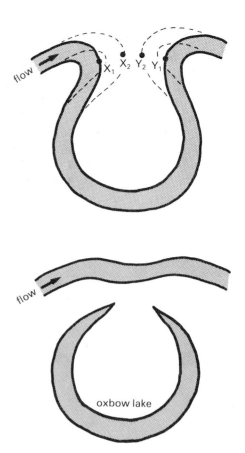

Fig. 5.31 Formation of an oxbow lake

The more curved a meander becomes, the greater the tendency it has to erode laterally back on itself. Fig. 5.31 shows this for a very well developed meander. In the diagram x_1 and y_1 are points of great lateral erosion/undercutting which in time will come to occupy positions x_2 and y_2. Eventually they meet, a full break develops and the stream abandons the meander loop as an **oxbow lake** or **cut-off**; this will slowly dry up.

> **21** Refer to meander 2 in Fig. 5.30.
> a) Where would you expect a future cut-off to develop in this meander? Draw a diagram to represent this.
> b) Such developments are quite common in rivers in their natural state. Why is it unlikely that such changes will occur here?

When a river reaches the sea its velocity may be suddenly checked, allowing the sediment load it is carrying to be deposited or redistributed by the sea. If the build-up of deposits is greater than the removal by the sea a **delta** will form. Deltas are not usually found on

River Processes and Associated Landforms

the coast of the British Isles as tidal scouring is active enough to prevent heavy deposition in the estuaries. However, deltas in a variety of shapes and sizes are common enough features throughout the world.

Fig. 5.32 shows what a typical delta might be composed of. The main stream may branch into a number of **distributaries** which may frequently change course, thereby affecting the shape of the delta. Distributaries deposit material and levees may build up; marshes develop with the gradual silting up of the zone between the distributaries, though some lagoons may remain. As the delta grows in size, a small coastal plain may also develop.

Fig. 5.33 is a sketch map of the Mississippi which we referred to above (see p. 113). The enormous load it carries to the sea is the result of erosion in its vast drainage basin; the delta has been developing steadily over the past 5000 years. Because of its present shape it is known as a **bird's foot** delta, whilst that of the Nile is described as an **arcuate** (arc-shaped) delta.

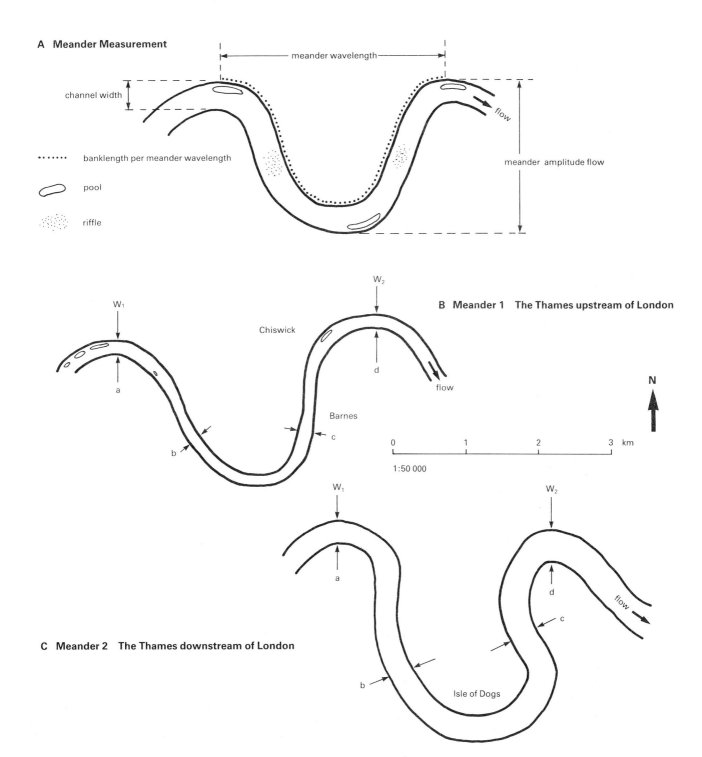

Fig. **5.30** Meander measurements and two meanders on the River Thames (see exercise 20)

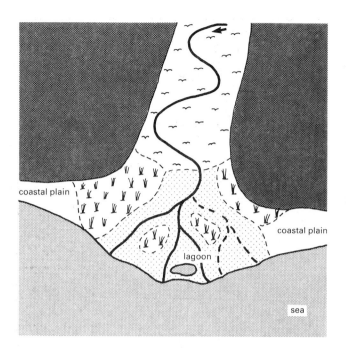

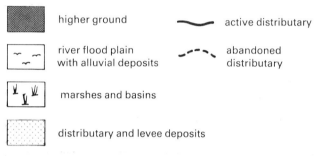

Fig. 5.32 Features of a delta

22 **a)** Draw a copy of Fig. 5.33 and using Fig. 5.32 as a guide, try to identify the various zones that have been marked by pecked (dashed) lines. Annotate your map.
b) From your atlas draw a sketch map of the Nile delta, annotating it as fully as you can.
c) In what ways, do you think, are deltas important to people, and what are the main hazards they present?

Changes in the River System with Time

We noted earlier how the headwaters of one river system might, by headward erosion, eventually 'capture' those of a neighbouring system. Fig. 5.34 shows how this might take place. In diagram A there are two original consequent streams, 1 and 2, in an area similar to that depicted in Fig. 5.3; stream 1a is a subsequent stream of 1. Since 1a is flowing across weaker rock, it can erode headwards more vigorously and its source steadily retreats along the pecked line; eventually it reaches point X. The steeper gradient of 1a will attract and divert the headwaters of 2. The effects of this are a) to leave a well defined **elbow of capture** at the

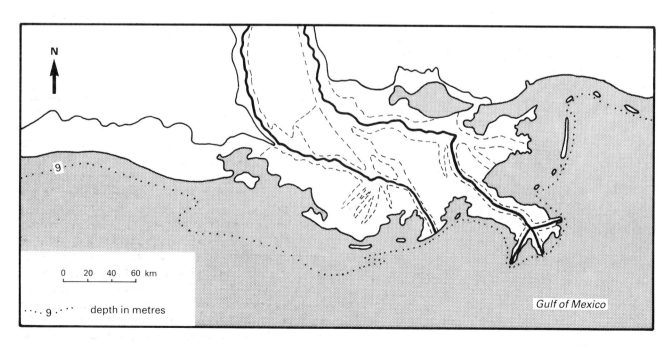

Fig. 5.33 The delta of the Mississippi

point of diversion, though in time lateral erosion may make this less distinct; b) stream 2 becomes a **misfit stream**, i.e. a stream which has a much smaller discharge than the size of its valley would indicate, having lost its headwaters; and c) a **wind gap** appears, part of stream 2's former valley where no stream is now flowing (see Fig. 5.34 B).

The River Rheidol in Mid Wales provides a good example of river capture (Fig. 5.35). The original headwaters of the Teifi rose on the slopes of Plynlimon and flowed south to enter the sea near Cardigan. At that time the Rheidol was only a small stream but was steadily eroding headwards along a line of weakness in the rocks. Eventually the source of the Rheidol met the Teifi at Devil's Bridge, at an altitude of about 250 m above present sea level. Because the distance to the sea along the Rheidol was considerably less than along the Teifi, the gradient was much steeper and the Teifi's headwaters were diverted. Immediately, these headwaters started eroding vertically at a very rapid rate, cutting a very deep and distinctive valley into the former valley. Today the waters of the Rheidol descend nearly 550 m in 45 km, representing one of the swiftest flowing rivers in Britain.

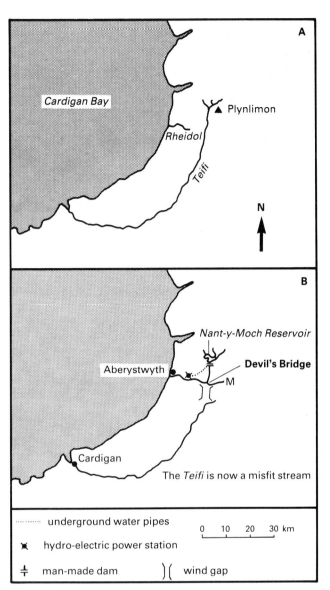

Fig. **5.35** The capture of the Teifi headwaters by the Rheidol at Devil's Bridge

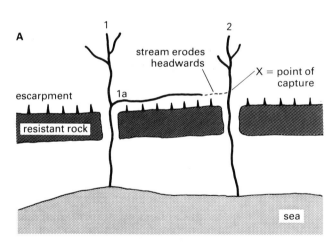

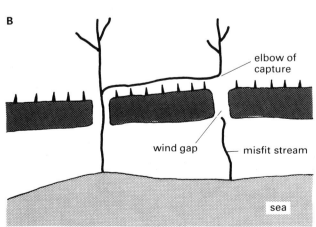

Fig. **5.34** River capture

Fig. 5.36 shows how extensive this vertical erosion has been, the house giving some idea of scale. The elbow of capture is easily picked out on Fig. 5.35 B, leaving a wind gap and the misfit Teifi to the south. The deep gorge that formed attracted other streams and the waterfalls in Fig. 5.25 A and B, the Mynach Falls, are on the stream marked M in Fig. 5.35 B, now a tributary of the Rheidol.

Such downcutting of an established stream into its own valley, for whatever reason, is known as **rejuvenation**. In the case of the Rheidol the effects were not confined to the gorge at Devil's Bridge. The pot holes shown in Fig. 5.20, for example, are only 2 km upstream from the point of capture. A further kilometre upstream can be found a dramatic illustration of the effects of rejuvenation. When downcutting is very rapid, even meanders do not have time to migrate

Map Extract 1 — Goring Gap

1:25 000 Second Series Crown Copyright Reserved

ROADS AND PATHS
Not necessarily rights of way

- M1 or A6(M) — Motorway
- A 31(T) — Trunk road
- A 35 — Main road
- B 3074 — Secondary road
- A 35 — Dual carriageway
- Road generally more than 4m wide
- Road generally less than 4m wide
- Other road, drive or track

Unfenced roads and tracks are shown by pecked lines

- Path

PUBLIC RIGHTS OF WAY (Not applicable to Scotland)
- Public paths { Footpath / Bridleway }
- Road used as a public path

Public rights of way indicated by these symbols have been derived from Definitive Maps as amended by later enactments or instruments held by Ordnance Survey on 1st Jun 1981 and are shown subject to the limitations imposed by the scale of mapping
The representation on this map of any other road, track or path is no evidence of the existence of a right of way

BOUNDARIES As notified to 1-3-77
- County (England and Wales) Region or Islands Area (Scotland)
- District
- London Borough
- Civil Parish (England)* Community (Wales)
- Constituency (County, Borough or Burgh)

Coincident boundaries are shown by the first appropriate symbol opposite

*For Ordnance Survey purposes County Boundary is deemed to be the limit of the parish structure whether or not a parish area adjoins

SYMBOLS
- Church { with tower / with spire / without tower or spire } or chapel
- Y — Glasshouse, Youth hostel
- Bus or coach station
- Lighthouse, lightship, beacon
- △ Triangulation station
- Triangulation point on { church, chapel, lighthouse, beacon, building & chimney }
- ° BP, BS — Boundary Post, Stone
- T, A, R — Telephone, public, AA, RAC
- P, MP, MS — Post office, Mile Post, Stone

- VILLA — Roman antiquity (AD 43 to AD 420)
- Castle — Other antiquities
- Site of antiquity
- ⚔ 1066 — Site of battle (with date)
- Gravel, sand pit
- Disused pit or quarry
- Chalk pit, clay pit or quarry
- Refuse or slag heap
- Sloping masonry
- W, Spr — Well, Spring

- Water
- Sand, sand & shingle
- Mud
- NT — National Trust always open
- NT — National Trust opening restricted
- NTS — National Trust for Scotland
- Electricity transmission line — pylon, pole

RAILWAYS
- Multiple track } Standard gauge
- Single track
- Narrow gauge
- Siding
- Cutting
- Embankment
- Tunnel
- Road over & under
- Level crossing, station

Information not available in uncoloured areas

DANGER AREA — MOD ranges in the area Danger! Observe warning notices

VEGETATION
Limits of vegetation are defined by positioning of the symbols but may be delineated also by pecks or dots

- Coniferous trees
- Non-coniferous trees
- Coppice
- Orchard
- Scrub
- Bracken, rough grassland — In some areas bracken (σ) and rough grassland (……) are shown separately } Shown collectively as rough grassland on some sheets
- Heath
- Reeds
- Marsh
- Saltings

HEIGHTS AND ROCK FEATURES
- 50 } Determined by { ground survey / air survey }
- 285·

Surface heights are to the nearest metre above mean sea level. Heights shown close to a triangulation pillar refer to the station height at ground level and not necessarily to the summit

Vertical face — Loose rock, Boulders, Outcrop, Scree

Contours are at 5 metres vertical interval

This map covers areas studied in depth in the text.
It is the most recent available at the time of publication.

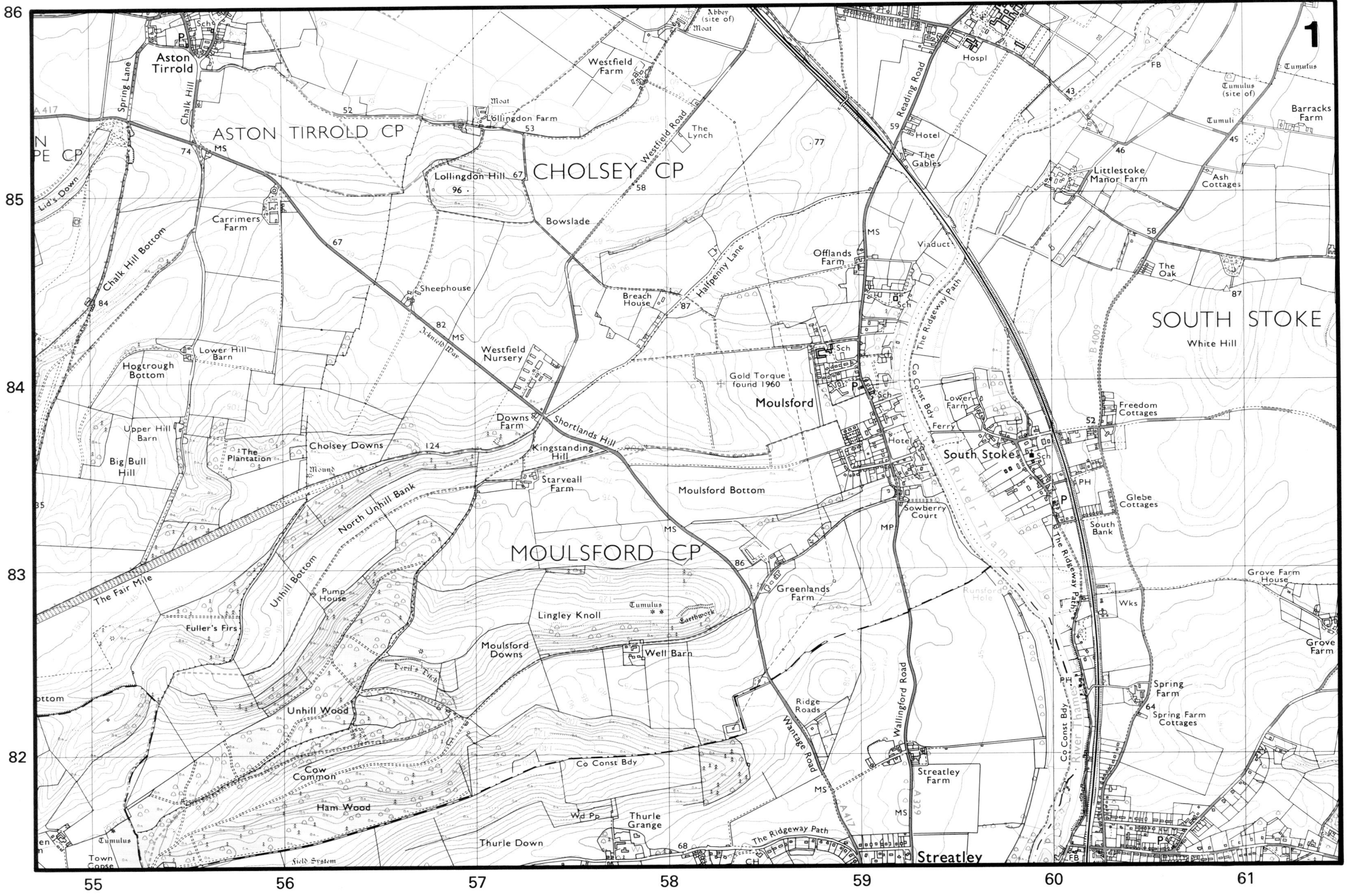
1
Aston Tirrold
ASTON TIRROLD CP
CHOLSEY CP
MOULSFORD CP
SOUTH STOKE
Westfield Farm
Lollingdon Farm
Lollingdon Hill
Bowslade
Westfield Road
The Lynch
Carrimers Farm
Chalk Hill Bottom
Spring Lane
Chalk Hill
Lid's Down
Sheephouse
Halfpenny Lane
Breach House
Westfield Nursery
Lower Hill Barn
Hogtrough Bottom
Upper Hill Barn
Big Bull Hill
The Plantation
Cholsey Downs
Downs Farm
Shortlands Hill
Kingstanding Hill
Starveall Farm
North Unhill Bank
Unhill Bottom
The Fair Mile
Pump House
Fuller's Firs
Lingley Knoll
Moulsford Downs
Well Barn
Unhill Wood
Cow Common
Ham Wood
Co Const Bdy
Thurle Grange
Thurle Down
Moulsford Bottom
Gold Torque found 1960
Moulsford
Offlands Farm
Reading Road
Hospl
Hotel
The Gables
Viaduct
The Ridgeway Path
Lower Farm
Ferry
South Stoke
River Thames
Sowberry Court
Greenlands Farm
Ridge Roads
Wantage Road
Wallingford Road
Streatley Farm
Streatley
Littlestoke Manor Farm
Ash Cottages
Barracks Farm
The Oak
White Hill
Freedom Cottages
Glebe Cottages
South Bank
Grove Farm House
Grove Farm
Wks
Spring Farm
Spring Farm Cottages
Tumulus
86
85
84
83
82
55
56
57
58
59
60
61

Map Extract 2 — Aberystwyth

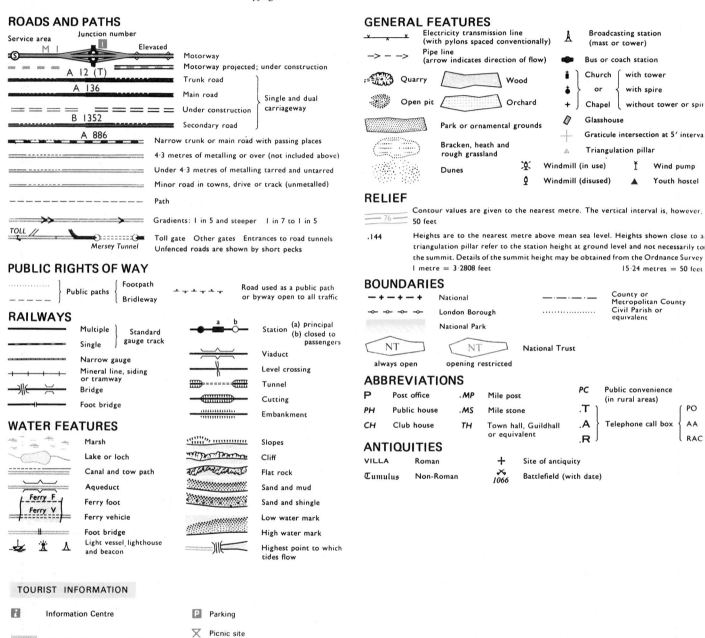

1:50 000 First Series Crown Copyright Reserved

This map covers areas studied in depth in the text.
It is the most recent available at the time of publication.

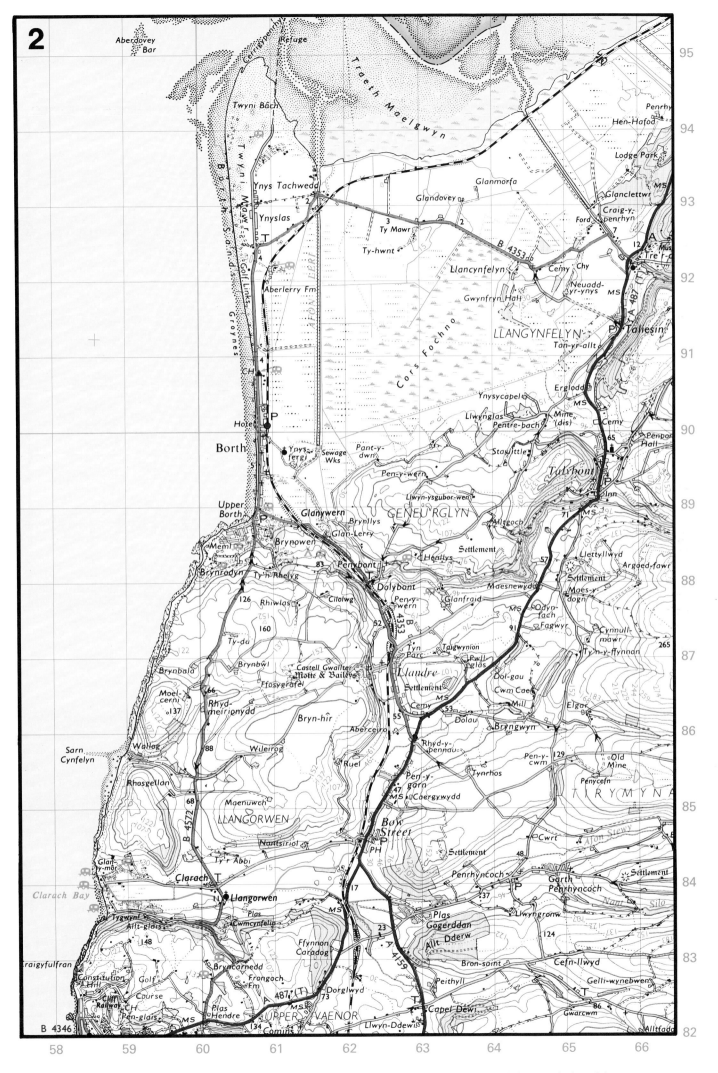

Reproduced from the Ordnance Survey Map 1:50 000 Landranger Series Sheet 135 'Aberystwyth' with the permission of the Controller of Her Majesty's Stationery Office, Crown Copyright reserved.

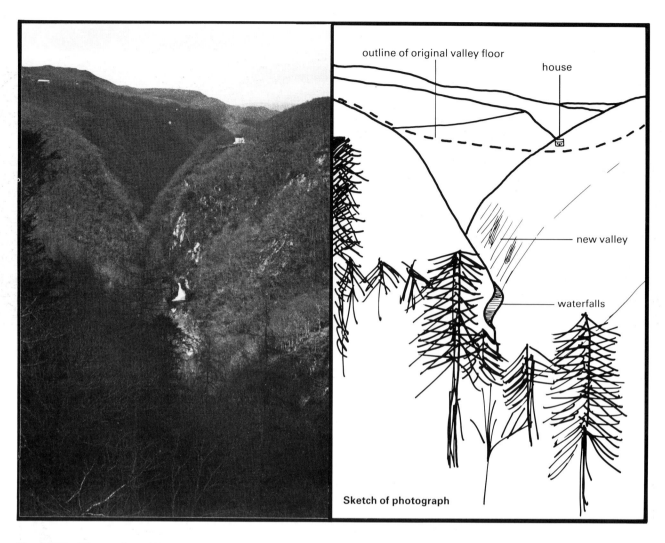

Fig. 5.36 The valley of the Rheidol looking north from Devil's Bridge

laterally as normal; the effect of this is to produce an **incised meander**, i.e. a meander flowing in a gorge instead of across a flood plain. Fig. 5.37 shows such a feature. Photograph A is taken looking across the old valley from one bluff line to another, whilst photograph B is taken looking south down the old valley and shows the gentle cross-profile of the former valley into which the incised meander has cut.

23 a) Draw a sketch of photograph A *or* B.
b) On it mark the following features:
  i) the abandoned flood plain of the meander core,
  ii) the former valley bluff line,
  iii) the present river and active valley,
  iv) the approximate angles (in degrees) of the valley sides of the incised meander.

Rejuvenation also follows from events other than river capture. It most commonly occurs when there is a relative fall in sea level, the effects of which are shown in Fig. 5.38. A **knick point** develops where the new long profile starts to cut downwards into the old, thus marking a distinct change of gradient. By headward erosion the knick point gradually retreats up the river valley. The abandoned flood plain of the former valley is left as a **river terrace** (B). In time, as the new flood plain broadens, river terraces may become very fragmented and may disappear altogether as a result of erosion, undercutting and mass wasting (C). Terraces may often be covered with old river deposits. Diagram D shows the effects of a fall in sea level on the river's long profile, with profile 3 showing successive positions of the knick point during its retreat. If a number of falls in sea level occur, then a river may have several knick points and several sets of terraces. Once the rejuvenation process has started, the river's long profile will continue to adjust, with the knick point retreating no matter what further changes occur (even a *rise* in sea level).

Fig. **5.37** Incised meander on the River Rheidol

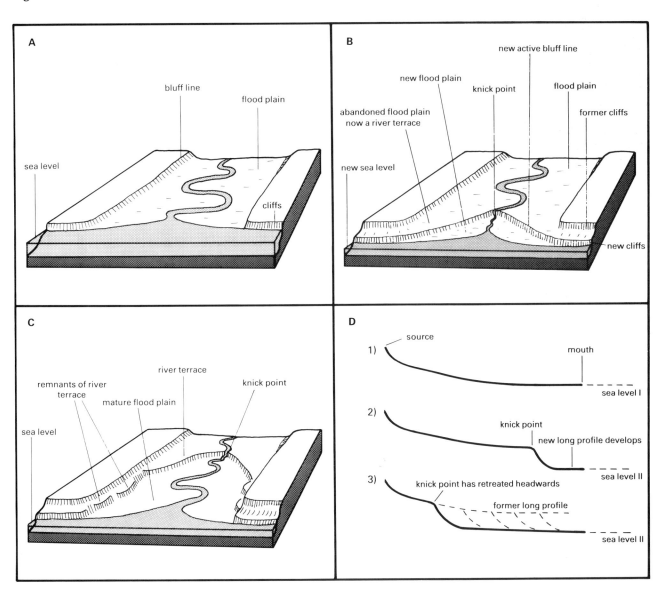

Fig. **5.38** The development of river terraces

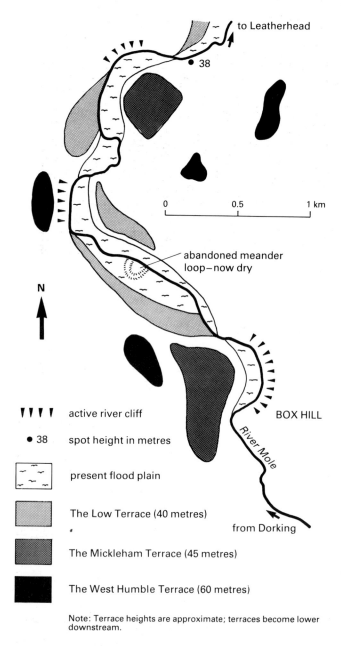

Fig. **5.39** Terraces of the River Mole between Dorking and Leatherhead

24 Locate the stretch of river shown in Fig. 5.39 in your atlas.

a) Using Fig. 5.39, draw a representative cross-profile of the Mole valley, showing how the valley might look if all the terraces were present along one section. Use a vertical scale of 1 cm to 25 m and calculate the vertical exaggeration of your profile.

b) Try to estimate the approximate width of each terrace from the fragments shown. Does the evidence of Fig. 5.39 suggest that the older the terrace the more fragmented it becomes? (Remember that the oldest terraces will be the highest above the present flood plain.) Why do you think some terraces are wider than others?

c) Using map extract 1, draw a cross-profile along northing 84 from eastings 58 to 61. Use the same vertical scale as your profile for the River Mole. The horizontal scales are the same. Mark on your profile the location of the river terraces using the information on the geological map.

d) Compare the two profiles you have drawn. Is there any justification for the suggestion that some of the terraces shown might have developed from the same falls in sea level?

Fig. 5.39 also shows examples of other river features, notably river cliffs and an abandoned meander loop. The river cliffs are at present being laterally undercut and are very distinctive. In fact the Box Hill river cliff, cut into chalk, has an angle of 45° and is nearly 120 m high.

The River Mole between Dorking and Leatherhead in Surrey shows a number of features which result from rejuvenation. Several sea level changes have led to the development of a series of river terraces above the present-day flood plain (Fig. 5.39). The figures in brackets on the key to Fig. 5.39 represent the approximate height of each terrace above present sea level. The Mole is a tributary of the Thames and it is interesting to note that the Thames also has well defined river terraces. Some of these are shown on the geological map relating to map extract 1 (see p. 26).

Man's Influence on River Systems

We have examined in some detail the processes and relationships that exist in a natural river system. However, such is the influence of man today that few systems, certainly in Britain, can be regarded as entirely natural. The industrial and domestic usefulness of water and the potential damage that flooding can cause provide two reasons for people wanting to modify the drainage basin.

Fig. **5.40** The valley of the River Findhorn, Scotland

A major example is the construction of dams to hold back and store water. In the case of the Upper Rheidol in Fig. 5.35 B, the Nant-y-Moch Reservoir has been created behind a 52 m high dam. It has a capacity to store 25 874 000 m^3 of water. This water is released through specially constructed underground pipes to the hydro-electric power station in the Lower Rheidol. The result is that the Rheidol headwaters now bypass much of the upper valley, leaving the stretch of river we examined above, featuring the Devil's Bridge gorge, waterfalls, pot holes and incised meanders, much less active. The Rheidol here has become an artificial misfit! This helps to explain why the pot holes in Fig. 5.20 should be so clearly visible.

However, much of man's influence on the natural process of drainage has been on a rather less grand scale than the construction of dams.

25 Consider what effects the following are likely to have on the natural drainage of a region:
a) the building of a new town,
b) an arable farmer laying a network of drainage pipes and ditches in his fields,
c) a landowner planting an extensive new forest.
You may find it helpful to refer back to Fig. 5.12.

26 The network of canals in Britain crosses many watersheds. How did the engineers overcome the problem of ensuring an adequate water level in the canals at these high points?

27 From map extract 2, consider how people have appeared to modify the drainage to the north of northing 90. Explain why this was done.

Workshop

1 Study Fig. 5.40 and draw a sketch of the photograph. Identify as many features resulting from fluvial activity as you can and annotate your sketch accordingly.

2 Study closely the pattern of drainage shown in Fig. 5.41. At what stage in its development does this river system appear to be? (Refer back to Fig. 5.3.) Consider the changes that have probably occurred to the system and the evidence you would look for *in the field* to confirm such changes.

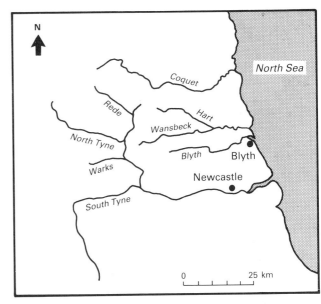

Fig. **5.41** Drainage in North East England

3 Review the chapter and list the various ways that river gorges may be formed.

4 Using map extract 1, suggest how the siting of settlements and communication lines in the valley of the Thames might have been influenced by the cross-profile form of the valley itself.

Glacial Processes and Associated Landforms 6

Over a long period of time, landscape evolution by running water depends on a region's climate remaining relatively unchanged, since processes and therefore landforms, vary between climates. It is now clear that the pattern of the world's climates existing today has not always been the same.

In geological terms, your life so far has been extremely short, but even so you will probably have been aware of some changes in climate as distinct from weather (refer back to Chapter 3, p. 43). Some summers may be described as 'good', i.e. hotter and drier than average (the summers of 1976 and 1983 in Britain were good examples), whilst some winters tend to be 'hard', i.e. colder and snowier than average (you will probably remember December 1981, your parents will remember the winter of 1963). Such changes represent variations from the normal pattern of climate and suggest that it is in a constant state of slow change. These changes are referred to as **climatic fluctuations**.

Short term fluctuations, say those over a period of decades or centuries, are best studied using written historical records such as personal diaries or, more recently, accurate weather records. These can then be compared with any natural evidence, for example that of tree rings. It is possible to take samples from the trunks of living trees to study the width of tree rings; although such analysis is very complicated, a wide ring in a tree in a cold climate generally indicates a warm year, whilst a wide ring in a tree in a dry climate indicates a wet year.

The following extract is from the diary of Gilbert White, a famous naturalist, writing in January 1776 about a remarkable frost. It had begun on January 7th.

'On the 27th much snow fell all day, and in the evening the frost became very intense. At South Lambeth for the following nights the thermometer fell to 11°, 7°, 6°, 6°, and at Selborne (Hampshire) to 7°, 6°, 10°; and on the 31st of January, just before sunrise, with rime on the trees and on the tube of the glass, the quicksilver sank exactly to zero, being 32° below the freezing point . . . A most unusual degree of cold this, for the south of England . . . The Thames was at once so frozen over, both above and below bridge that crowds ran about on the ice . . . What (snow) had fallen on the roofs was so perfectly dry that from first to last it lay 26 days on the houses in the city – a longer time than could be remembered by the oldest housekeepers living.'

> **1** Study the extract carefully; note the temperatures are in degrees Fahrenheit.
> a) Compare these temperatures with those for a recent January period, such as that in Table 3.1. If your school keeps its own weather log you should consult that; if not refer to the weather log published by the Meteorological Office.
> b) Summarise the evidence in the extract for the severity of the cold period.

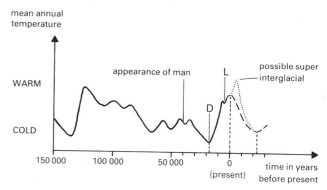

D: The Devensian Glacial

L: The Loch Lomond Re-advance

— — — likely future change in temperature

.... possible future change in temperature

Fig. **6.1** Temperature changes in Britain

Several cold periods over a number of years may be indicative of a definite short term climatic fluctuation. In fact between 1600 and 1814, the Thames in London froze over twenty times and in the winter of 1683–4 it froze for two months with the ice reaching a thickness of nearly 30 cm. However, by itself the freezing of the Thames is not a perfect guide to climatic change.

> **2** Suggest reasons, other than climate, why the Thames would be unlikely to freeze today even during a cold spell.

Glacial Processes and Associated Landforms

For longer term climatic fluctuations, over tens of thousands of years, historical records are not available. Scientists must therefore refer to other evidence, like that contained in the deposits which have accumulated on the sea floors. Such evidence has enabled a picture of the temperature changes for Britain during the last 150 000 years to be built up, as shown in Fig. 6.1. The troughs on the graph represent cold periods, the coldest of which are known as **glacials** (Ice Ages) when the polar ice caps expanded significantly; the high peaks represent **interglacials** when the ice caps became much smaller. It has been established that in the last 2.1 million years there have been at least 21 glacials – this is the geological period known as the **Pleistocene**. The most recent glacial, the **Devensian**, can be seen on the graph. You can also see that the last few thousand years have been spent in a relatively warm interglacial. The graph shows that 'hiccups' can occur during climatic change, like that of the Loch Lomond Re-advance. Most forecasts indicate that the Earth's climate is slowly cooling to another glacial period (it has cooled by as much as 2°C over the last 700 years). However, some scientists believe that before this glacial occurs there will be a 'super interglacial', with temperatures much higher than in other interglacials.

3 Study Fig. 6.1.
a) How many years ago was the Devensian glacial?
b) How many years will it be before the next predicted glacial maximum is reached?
c) Do you think that people will be able to survive the next glacial? Justify your answer.

A number of factors are believed to have an influence on climate:
a) Over long periods, the Earth's orbit around the Sun is known to change – it may be elliptical (oval) or near circular. The more elliptical the orbit, the cooler the climate. The axial tilt of the Earth to its orbital path, at present $23\frac{1}{2}°$, is also known to change.
b) The energy emitted from the Sun – solar radiation – periodically varies and when this is high, climatic disturbances on Earth generally follow. Such variations appear to follow a fairly regular pattern.
c) Changes in the gaseous composition of the atmosphere also bring about climatic change. It is known that as the amount of carbon dioxide increases, so average temperatures tend to rise, since the gas helps to prevent solar radiation from escaping back into space. This is known as the **greenhouse effect**. The burning of fossil fuels (coal, oil, gas, etc.) releases carbon dioxide into the atmosphere and encourages this effect.
d) The amount of dust in the atmosphere may have an influence on climate. Dust is given off by volcanic activity and the greater the levels of dust, the more likely it is that the Sun's rays will be reflected back into space, thereby cooling the climate. Fig. 6.2 shows the relative dust levels in the atmosphere since 1500; it shows that in the early 16th century and mid 20th century dust levels were extremely low.

4 a) Do you consider that the dust level variations in Fig. 6.2 help to explain the great frosts of London in the 17th and 18th centuries, and in particular the two mentioned above, 1683–4 and 1776?
b) Is it possible that the eruption of Mount St. Helens examined in Chapter 1 will have/has had any effects on climate?
c) By reference to the possible causes of glacials, can you suggest a reason why a super interglacial should be predicted in Fig. 6.1?

Fig. **6.2** Record of volcanic dust in the atmosphere of the northern hemisphere since 1500

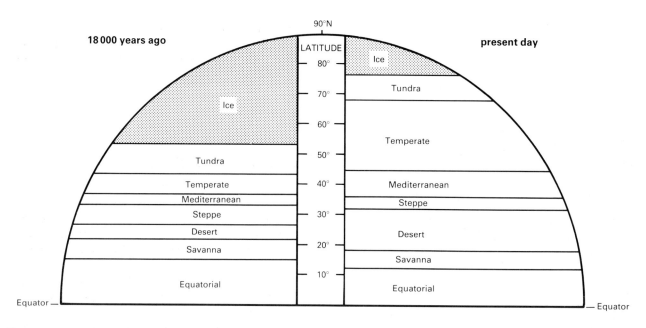

Fig. 6.3 Climatic zones of the northern hemisphere, past and present

During glacials, temperatures drop and much more of the Earth's *fresh* water becomes stored in the form of snow and ice, particularly in the polar regions (today 75% of the total is stored in this way). As a result, the accumulation of snow and ice begins to exceed the rates of wastage and **ablation** (melting). Polar ice caps and ice sheets therefore expand and 'push' the Earth's climatic zones towards the Equator. Fig. 6.3 shows what the distribution of these zones would have been like during the Devensian glacial 18 000 years ago, compared with today. However, this glacial was not the coldest nor the longest in the Pleistocene period, when at their maximum extent the ice sheets covered 30% of the Earth's surface, compared with only 11% today.

5 Study Fig. 6.3.
a) What was Britain's position in relation to the climatic zones 18 000 years ago? (Use an atlas to find Britain's latitude.)
b) Refer to Chapter 4 and try to describe what the climate of your home region would have been like at that time.

Fig. 6.4 shows the actual limits of ice advance over England and Wales during the last glacial; the cities are

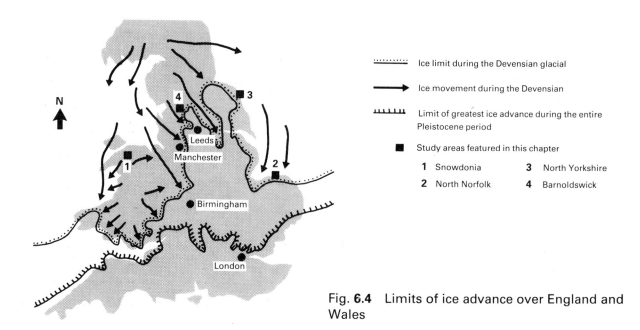

Fig. 6.4 Limits of ice advance over England and Wales

Glacial Processes and Associated Landforms

shown for reference purposes. Also marked is the limit of the greatest ice advance during the entire Pleistocene period. You can see that most of Britain has at some stage experienced glaciation.

The Formation of Ice

As the climate of a region becomes steadily cooler, so patches of snow which fell during the winter remain on the ground for progressively longer periods into the summer. Eventually they may become permanent and when this happens the process of ice formation can begin.

Freshly fallen snow traps many air spaces within it and so has a density of only 0.1 g/cm^3 compared with water which has a density of 1.0 g/cm^3. The weight of further snowfalls will compress previous ones, thus helping to force out the air and increase the density. When temperatures rise above 0°C, some melting of surface snow will occur. Drips of meltwater will slowly seep down through the lower snow layers to fill some of the air spaces before refreezing. The effect of these two processes of compression and meltwater seepage is to increase the snow density to between 0.4 g/cm^3 and 0.8 g/cm^3. This produces an intermediate form of ice known as **firn** or **neve** which in time will be transformed to ice itself with a density of 0.9 g/cm^3. An average time for ice formation might be about 70 to 100 years, but in areas where snowfalls are heavy and temperatures regularly fluctuate around 0°C, this period may be as short as 4 or 5 years. By contrast, in areas such as the Antarctic where snow tends to be 'dry' and temperatures well below 0°C, the process may take much longer, 1000 years or more.

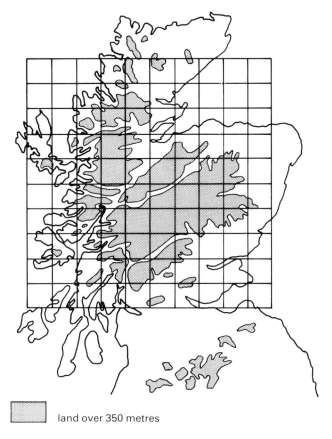

land over 350 metres

Fig. **6.5 (a)**   Map of highland in North Scotland

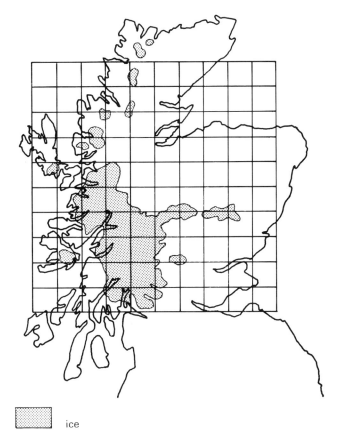

ice

Fig. **6.5 (b)**   Map showing the extent of the Loch Lomond Re-advance

00	01	02	03	04	05	06	07	08	09
10	11	12	13	14	15	16	17	18	19
20	21	22	23	24	25	26	27	28	29
30	31	32	33	34	35	36	37	38	39
40	41	42	43	44	45	46	47	48	49
50	51	52	53	54	55	56	57	58	59
60	61	62	63	64	65	66	67	68	69
70	71	72	73	74	75	76	77	78	79
80	81	82	83	84	85	86	87	88	89
90	91	92	93	94	95	96	97	98	99

Fig. **6.5 (c)**   Numbering of grid squares shown in maps (a) and (b)

Permanent snow and ice may occur in a number of forms. **Ice sheets** cover vast continental land surfaces, whilst **ice caps** are much more localised in extent. Tongues of ice or **glaciers** may extend from both of these. A cover of ice at sea creates an **ice shelf** from which ice fragments or **icebergs** may break loose. Today 80% of the Earth's permanent snow and ice is found in the Antarctic, covering 14 million km².

> **6 a)** Explain briefly why the Antarctic's climate is so cold.
> **b)** Use your atlas to draw a map of Antarctica. On it mark, and where possible name, ice sheets, ice caps, ice shelves and the limits of drifting ice.

One important influence on the coldness of the Antarctic climate is altitude; large areas are over 3600 m high. Altitude also has a great effect on ice development in temperate latitudes, as in the Alps and Himalayas today. (Read p. 50 again to see why this is so.) It is therefore always in upland regions that the early signs of a glacial can be detected, with the steady increase in the amount of permanent ice. Such was the case in Britain with the Loch Lomond Re-advance shown in Fig. 6.1. Although this was only a temporary cooling interlude in a period of overall climatic warming, it did lead to substantial development of ice, as shown in Fig. 6.5. There is obviously a similarity between maps (a) and (b), though the match is not a perfect one. An exercise can be undertaken to show how well the two maps correspond – it is best expressed in the testing of a simple hypothesis.

> **7** Hypothesis: *that ice development was related to altitude in the Loch Lomond Re-advance*. In order to test this hypothesis you should adopt the following procedure:
>
> **a)** For a sample number of points on the map, you will need to establish i) whether or not ice developed, and ii) whether the point is highland or lowland.
>
> **b)** Select 20 sample points at random; this removes personal choice and helps to keep the exercise objective. The grid of 100 squares overlying both maps (a) and (b) will help you do this. Each square is identified by a number, 00 to 99 in Fig. 6.5 (c), and to select a square, a two-digit random number must be chosen. Follow the same procedure as described in Chapter 2 (p. 28) using Table 2.1. Use the south-west corner of each selected square for your point recordings.
>
> **c)** Enter your results for the 20 randomly selected points on a copy of the **matrix** (table) shown in Table 6.1 (a). Simply place a tick for each sample point in the appropriate box on the matrix. You may have to disregard some points you select, such as those at sea. Since these represent your own sample observations the matrix is known as an **observed frequency** (f_o) matrix.
>
> **d)** Complete the 'totals' boxes in the matrix; the combined glaciated/unglaciated points total should be the same as the combined highland/lowland points total.
>
> **e)** You must now calculate what the distribution of points within the matrix would have been with those same totals, assuming that *no* link between ice development and altitude existed. Such a 'no link' distribution is known as an **expected frequency** (f_e). Table 6.1 (b) gives an example 'totals distribution' which might have been observed. In the manner shown, calculate the expected frequencies for your own totals (which will probably be different from the example shown). Enter the results on an expected frequency matrix.
>
> **f)** You should now compare the two matrices. Since the expected frequency matrix provides you with a 'no link' distribution for your sample points, the greater the numerical difference between the observed and expected frequencies, the stronger will be the relationship between ice development and altitude.
>
> **g)** On either matrix shade in those boxes where the observed frequency is greater than the expected. Interpret your result in terms of your original hypothesis. Thus, if you have shaded the 'highland glaciated' and 'lowland unglaciated' boxes, your hypothesis can be accepted.
>
> **h)** Had there been a perfect relationship between ice development and altitude in your sample, what observed frequency distribution would you have expected?

Columns Rows	Glaciated	Unglaciated	Totals
Highland			
Lowland			
Totals			20

Table **6.1 (a)**   Observed frequency matrix

Columns Rows	Glaciated	Unglaciated	Totals
Highland	I	II	12
Lowland	III	IV	8
Totals	7	13	20

The expected frequency (f_e) for box I is found by multiplying the row total by the column total and dividing by 20. The f_e of other boxes is found in a similar way.

Thus: $f_e \text{ I} = \dfrac{12 \times 7}{20} = 4.2$

$f_e \text{ II} = \dfrac{12 \times 13}{20} = 7.8$

$f_e \text{ III} = \dfrac{8 \times 7}{20} = 2.8$

$f_e \text{ IV} = \dfrac{8 \times 13}{20} = 5.2$

Table **6.1 (b)**   Example expected frequency matrix

The Effects of Ice Movement

Glacial Erosion

The effects of past glaciations are most obvious in upland regions such as those in Scotland. Ice, like running water, acts as an agent for erosion, transportation, and the deposition of weathered material.

Glaciation starts with the formation of the ice itself. Rock fragments, which are broken loose by **freeze-thaw weathering** (see Chapter 2), become embedded in the ice and provide the material or load which allows erosion to take place once the ice begins to move. A very rugged weathered surface is left above the snow and ice. Fig. 6.6 shows how such freeze-thaw weathering has attacked a small rock outcrop in the mountainous region of Cader Idris (shown in Fig. 6.37), even though there is no permanent snow and ice today. The large piles of scree in the background result from the same process.

Fig. **6.6**   Freeze-thaw weathering of a small rock outcrop in the Cader Idris region of Mid Wales

As the climate becomes colder and the thickness of the ice increases, so greater pressure is exerted on the ice below, eventually causing it to flow at its margins. Gravity will also be important, encouraging the ice to flow downslope; the steeper the slope, the faster the rate of flow. Assisting ice movement is lubrication by meltwater at the base of the ice (temperatures will not always be below 0°C at the base). **Basal slip** will take place. Speeds of flow will vary according to the conditions, but they range from about 50 m a year in the case of some Alpine glaciers, to several kilometres a year in the case of some Greenland glaciers.

Erosion of the bedrock takes place as the weathered load trapped in the ice is dragged along by ice movement. This leads to the **abrasion** of the bedrock surface and eventually to the grinding down of the rock fragments themselves into a 'rock flour'. By this process, extra fragments are always being added to the ice, having been broken loose from the bedrock below.

In order to confirm this, scientists undertook an experiment in the Breidhamerkurjokull Glacier in South East Iceland. They bolted two pieces of rock to the bedrock beneath the glacier – one of marble (metamorphosed limestone), the other of basalt (a more resistant volcanic rock). The rocks were left for three months during which time the glacier moved 9.5 m. The following results were obtained:

a) Both rocks were heavily scratched.
b) A corner of the basalt rock had broken off and disappeared.
c) The basalt had been lowered by an average of 1 mm.
d) The marble had been lowered by an average of 3 mm.

> **8 a)** Do these results uphold the view of glacial erosion described above? Justify your answer.
> **b)** What conclusion can you draw about the influence of rock resistance on the nature and rate of erosion?
> **c)** From the figures quoted, calculate the degree of landscape lowering that might occur in regions with rocks of similar resistance to those mentioned above, with ice flow lasting for 2000 years – the time span of the Loch Lomond Re-advance.

Fig. **6.7** Striations

Surface scratches are known as **striations** (striae) and may still be evident on rock surfaces once the ice has disappeared. Fig. 6.7 shows striations extending across the rock; they provide a good indication of the direction of ice flow.

Another way in which ice movement can lead to bedrock erosion is by the process of **ice plucking**. When ice freezes to a rock surface, the strength of its adherence may cause subsequent ice flow to pull rock fragments away. This occurs more generally on the sheltered sides of rock surfaces where ice abrasion is not so active.

The **roche moutonnée** is a good example of a landform which owes its shape to both ice abrasion and plucking. Fig. 6.8 shows an example in the Cader Idris region, a rock outcrop which has been slightly more resistant to ice erosion than the surrounding rock. Photographs A and B show the side and front views respectively of the roche moutonnée, whilst C is a representative diagram of a typical roche moutonnée, showing the characteristic features.

> **9** Study Fig. 6.8 closely. This feature is 63 m long, with the side that is visible 7.6 m high.
> **a)** Draw a sketch of photograph B and annotate it carefully, identifying all the relevant features.
> **b)** Mark on your sketch the direction of the ice flow which helped shape the landform.

Roches moutonnées are relatively small-scale features and invariably develop within larger ice-eroded features, like those shown in Fig. 6.9. These are **cirques** (also called **corries** or **cwms**) with their steep back and side walls producing an amphitheatre-like landform. Cirques extend into 'U' shaped valleys, a contrast with the 'V' shape produced by rivers in upland areas.

Glacial Processes and Associated Landforms

Fig. 6.8 A roche moutonnée

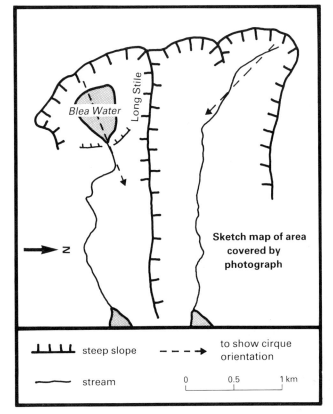

Fig. 6.9 Blea Water, Cumbria

We can develop some understanding of how these larger features may have been formed by examining actual glaciers today. Fig. 6.10 shows the long profile through a small cirque glacier in the Jotunheimen region of Norway. Scientists here were able to build up a picture of ice movement by digging tunnels into the glacier. They found a pattern of flow as shown in Fig. 6.10 A; the lines 1, 2, 3, etc. represent successive positions, at 10-year intervals, of dirt bands which accumulate on the ice surface during summer months when snowfall is at a minimum. Diagram B shows that the ice is moving in a rotational fashion within the cirque, with the ice at the cirque bottom flowing faster than that at the surface. Abrasion will cause the cirque

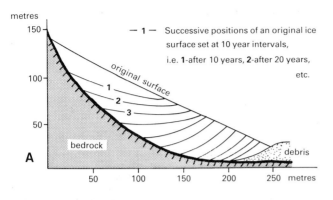

to deepen, leading eventually to glacial **overdeepening** which can produce 'reversed gradients', i.e. places where the ice is actually flowing upslope. You should notice the build-up of debris at the foot of the glacier; this was the load carried by the ice and dumped on melting. Notice also the steep back wall of the cirque; glacial plucking is very active here.

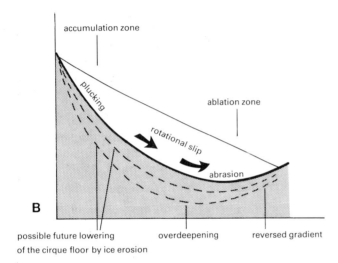

Fig. **6.10** Long profile of a cirque glacier in the Jotunheimen region of Norway

10 a) Locate the Jotunheimen region in your atlas. Compare the latitude and altitude of the region with those of the part of Scotland where the Loch Lomond Re-advance occurred. Which of these two factors is most likely to explain the presence of glaciers in this part of Norway today?
b) Refer to Fig. 6.10 A. How many years does it take ice to flow from the top of the cirque to the end of the glacier?
c) What do you understand by the terms: i) accumulation zone, ii) ablation zone? (Fig. 6.10 B.)
d) Refer to Fig. 6.9. Which of the two cirques shown has been overdeepened and contains reversed gradients? Justify your answer.

Fig. **6.11** Rhône Valley Glacier, Switzerland

Glacial Processes and Associated Landforms

The Rhône Glacier shown in Fig. 6.11 is a typical valley glacier, a long tongue of slowly moving ice. On the photograph notice the snowfields in the background which supply the glacier and notice too how the glacier has actually flowed below the snow-line (see Figs. 4.34 and 4.35). The downslope limit of the glacier is known as the **snout** and towards the snout the glacier becomes thinner and narrower. Notice the streams of meltwater flowing away from the snout.

Such glaciers generally flow along pre-existing lines of weakness in a landscape, often river valleys. As they do so, they reshape the landscape features. Fig. 6.12 shows the typical nature of valley glacier flow. Ice will tend to flow more rapidly at the glacier surface than at its base, A, (in contrast to cirque glaciers) and more rapidly in the centre of the glacier than at its sides, B. Freeze-thaw weathering of the upper valley slopes will produce rock fragments which will fall into the ice and be concentrated near the valley sides. Abrasion and plucking of the valley floor by the flowing ice will concentrate a load there. Therefore, where the ice and its load are in contact with the bedrock, friction and thus erosion will be greatest and flow slowest. Fig. 6.12 C contrasts the smoother ice-abraded valley sides with the more rugged slopes above the ice surface. With the ice protecting the valley sides and bottom from weathering, glacial erosion progressively transforms the valley to a deep trough with a characteristic U shape, as shown in diagram D.

> 11 Study Fig. 6.11 and read carefully through the preceding two paragraphs.
> a) Draw a sketch of the photograph in your exercise book.
> b) Annotate the sketch with all the features described or mentioned in the paragraphs, e.g. snowfields, snow-line, etc.

You will also notice in Fig. 6.11 that the glacier does not swing from side to side in the same way as running water. Because of this any spurs that may have existed are **truncated**, literally cut away from the valley side.

Fig. 6.13 shows the Barnard Glacier in Alaska and this confirms the direct nature of glacial flow; it also shows that networks of glaciers can exist. You will observe that the tributary glaciers are much smaller than the main glacier. This has consequences for landform development since the amount of downcutting is related to ice mass – the bigger the glacier, the deeper it will cut. The bottom of a main glacier will consequently be lower than that of a tributary glacier. When the ice eventually disappears, the tributary valley will be left suspended above the main valley as a **hanging valley.**

Rotational slip and overdeepening (observed in the cirque glacier) are not so common in valley glaciers.

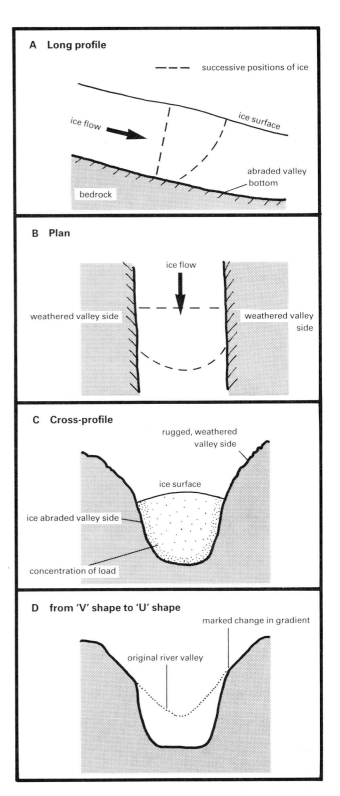

Fig. **6.12** Representative diagrams of a typical valley glacier

However when they do occur, a long profile like the one in Fig. 6.14 may develop, producing a series of rock steps. When the glacier has retreated, each step or **sill** will provide an ideal dam for a lake, which, because it is enclosed in a relatively straight valley, will be long and narrow. Such lakes are known as **ribbon** or **finger lakes**,

Fig. 6.13 The Barnard Glacier, Alaska

whilst the lakes left in overdeepened cirques are referred to as **tarns** (see Fig. 6.9). Fig. 6.15 shows a good example of a ribbon lake in a very distinctive trough valley. With the disappearance of the ice, post-glacial streams from hanging valleys may deposit enough material to divide a ribbon lake in two. It is worth noting that there is no limit to glacial overdeepening and ribbon lakes may be scoured out to a depth well below sea level.

When two trough valleys or cirques develop adjacent to one another, a knife-edged ridge or **arête** may develop. When several cirques develop around a mountain mass, a **pyramidal peak** may result; the Matterhorn in the Alps is the classic example. Both features generally show signs of heavy freeze-thaw weathering.

Fig. 6.16 is a block diagram of the main features in a glaciated upland region which have been mentioned so far.

> **12** Draw a copy of the diagram in your exercise book and on it identify the features a–h.

We can now build up a picture of the likely stages in the glaciation of an upland region in the northern hemisphere. Fig. 6.17 I shows a north–south cross-section through a typical unglaciated round-topped hill.

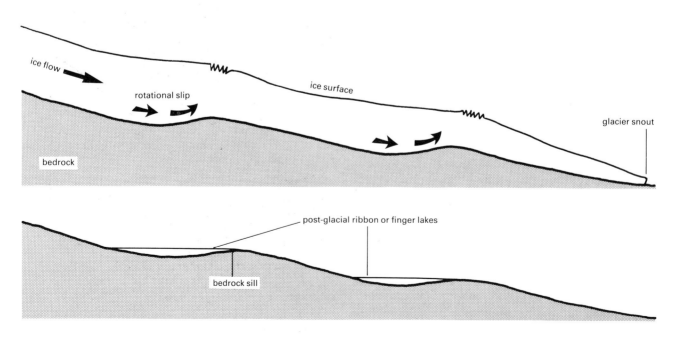

Fig. 6.14 Long profile of a trough valley during and after rotational slip ice abrasion

Glacial Processes and Associated Landforms

13 a) Study Fig. 6.17 carefully. The contrast between stages I and IV highlights the differences between a glaciated and unglaciated upland region. Describe them.

b) Can you suggest why the north side of the hill should have undergone glacial modification before the south side? (Fig. 3.17 provides a clue.)

c) Refer back to Fig. 6.9. What are the correct geomorphological terms for Blea Water and Long Stile? Contrast the direction faced by the overdeepened cirque occupied by Blea Water and the cirque on the right. The pecked lines identify the orientation of the cirques.

To confirm the idea that orientation will influence cirque development a network of cirques needs to be examined. A good example is available in the Snowdonia region of North Wales, as shown on Fig. 6.18. The steep slopes marking the rims of the cirques have been shown. Where rims of adjacent cirques coincide, a well defined arête is formed.

Fig. **6.15** Loch Einich, Scottish Highlands

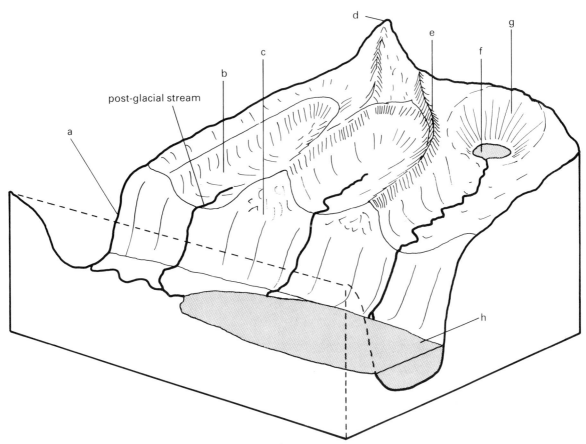

Fig. **6.16** Features of a glaciated upland

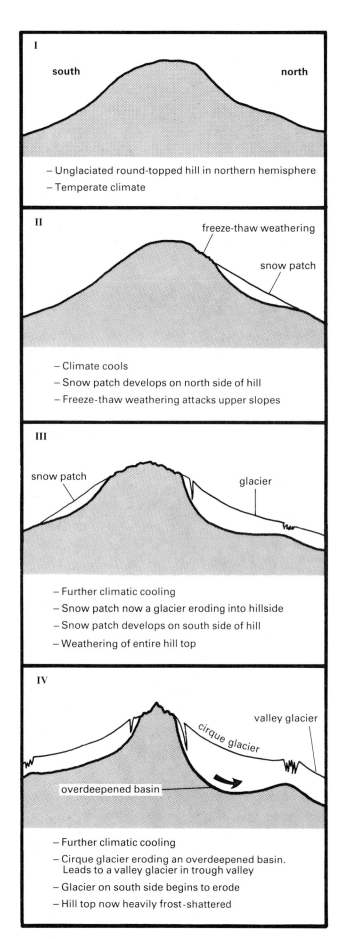

Fig. **6.17** Stages in the glaciation of a hill in the northern hemisphere

14 Study Fig. 6.18 carefully.
a) Using the method shown in Fig. 6.9, determine the orientation of each of the 14 cirques marked a–n. Each orientation should be the most appropriate of the eight main bearings, i.e. N, NE, E, SE, S, SW, W and NW. Fig. 6.18 C can be used as a guide.
b) Enter your results on a copy of the table (Fig. 6.18 B) and count up the number of cirques for each compass bearing.
c) Present your results in the form of a circular graph (Fig. 6.18 C). Using a suitable scale, draw a line down the centre of each sector of the graph, proportional to the number of cirques being represented.
d) What does this graph show and what conclusions can you draw?
e) For interest, post-glacial tarns have been marked. Does their location help to support the ideas first expressed in Fig. 6.17?

You should by now appreciate the importance of **aspect** (i.e. the orientation of slopes in relation to the Sun) in glacier development. The orientation of cirques and U shaped valleys provides a clear record of the direction of ice flow.

Glacial Deposition

We have seen how rock fragments become embedded in a glacier during its formation. As these are transported away downslope, other fragments are constantly being added by glacial erosion and by freeze-thaw weathering of rock faces above the glacier. A further study of Fig. 6.13 reveals distinctive lines on the glacier's surface. These are trains of rock debris known as **moraines**. As weathered rock falls away from the rock face it collects and embeds on the side of the glacier to form a **lateral moraine**. When a tributary glacier joins a main glacier a lateral moraine may be forced into the middle of the ice flow to create a **medial moraine**. Both types of moraine are identified on Fig. 6.19 which also shows how the rock fragments on the bottom of a small tributary glacier will become a band of debris within the main glacier; this is known as an **englacial moraine**. You will see from Fig. 6.19 that A is the shallower of the two tributary glaciers, since its resultant englacial moraine is not as deep as that of B. Moraines beneath a glacier are called **subglacial moraines**. As the glacier's mass increases, so does its load. This will be progressively deposited as the glacier reaches lower, warmer regions.

Glacial Processes and Associated Landforms

A Location of cirques in Snowdonia

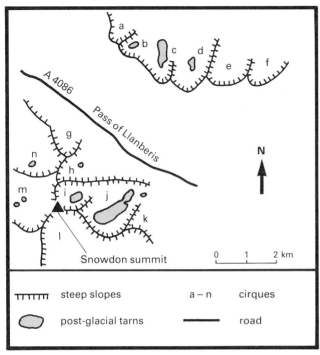

One of the influences on glacial deposition is the relative movement of the valley glacier's snout. The snout marks the point where ice supply from the glacier is balanced by ice wastage (ablation) through melting, evaporation, etc. The position of the snout may advance or retreat, reflecting short term changes in climate. However, you should remember that whether the snout is advancing or retreating, the glacial ice will continue to flow downslope. Fig. 6.20 shows how different climatic circumstances can affect the position of the snout. In diagram A the climate is stable and the snout remains static, but the stake has moved downslope and will eventually be dumped. In diagram B climatic cooling has caused the snout to advance as the snow and ice fields supplying the glacier grow; the stake may advance more rapidly downslope. In diagram C the snout is retreating as the climate gets warmer, but the stake continues to move forward (I and II), though eventually it is dumped (III).

When the snout is static, a build-up of deposits occurs as the ice continues to flow and transport its load. In time, this process leads to the formation of a long ridge which at the furthest limit of ice advance is known as a **terminal moraine**. However, if the snout remains static in a period of overall ice retreat (recession), the resultant ridge of deposits is known as a **recessional moraine**. Recessional moraines are therefore always located behind terminal moraines. It is difficult to generalise on the size of these features, but some terminal moraines may be 200 m high and several kilometres in length; mostly though they tend to be smaller than this.

When the snout advances the ice may push deposits into a ridge identical to the terminal and recessional moraines; this is a **push moraine** and is often combined with a terminal moraine.

During a period of climatic warming the snout retreats and the entire glacial load is usually dumped haphazardly over the valley floor. The collective term for these deposits is **boulder clay** or **ground moraine** ('till' is also used). Boulder clay deposits from an ice sheet cover very large areas, much of East Anglia for example. Since the ice would most probably have flowed over several bedrock types, the character of boulder clay is extremely variable. Irregular rock fragments are simply dumped in an unsorted matrix of rock flour (see p. 140), sands and clays, with the proportions of these constituents differing from place to place.

A detailed study has been made of the changing position of the snout of the Franz Joseph Glacier on South Island, New Zealand, one of several glaciers descending from the Southern Alps towards the west coast. Recorded speeds of glacial flow here range from 1.5 to 7 m per day. The positions of the snout between 1894 and 1967 are shown in Fig. 6.21. The snout was in retreat after 1967.

B

cirque	direction faced	cirque	direction faced
a		h	
b		i	
c		j	
d		k	
e		l	
f		m	
g		n	

C

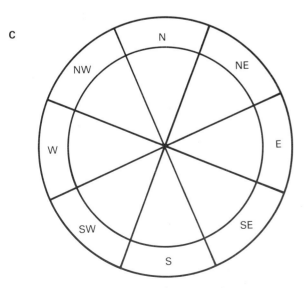

Fig. **6.18** To show the orientation of cirques in Snowdonia (see exercise 14)

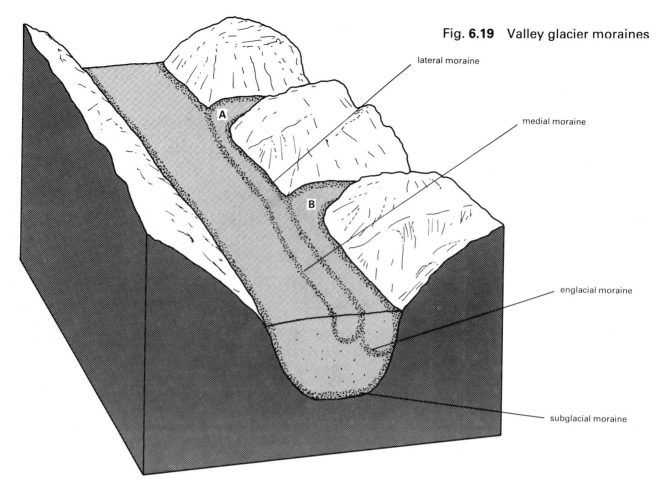

Fig. 6.19 Valley glacier moraines

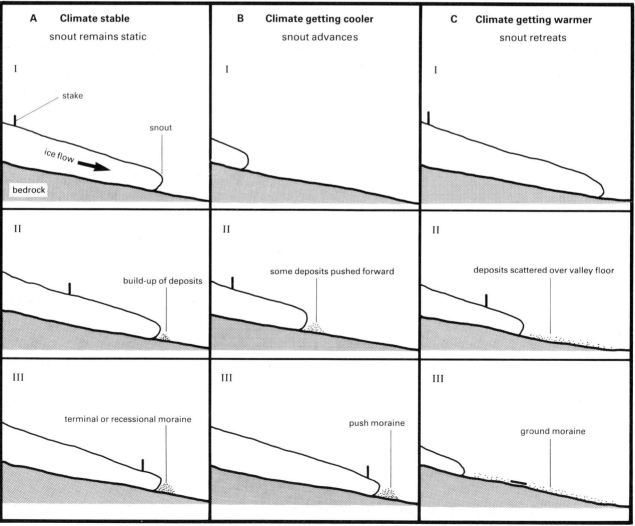

Fig. 6.20 Influence of climate on the position of a glacier's snout

Glacial Processes and Associated Landforms

15 Study Fig. 6.21 and locate the Southern Alps, New Zealand in your atlas. For reference, the Franz Joseph Glacier is 10 km to the north of Mount Tasman.

a) Copy the accompanying table into your exercise book (Table 6.2). Complete columns I–IV, using the scale on the sketch map for column III.

b) Which period showed the greatest average annual rate of change of the snout position?

Scientists believe that the advances of the Franz Joseph's snout are linked to the amount of winter precipitation on the snowfields *five years before*. The advance between 1965 and 1967 is thought to be related to the very high winter precipitation in 1962. Though significant, such a link is probably too simple to be applied to *all* glaciers. The nearby Fox glacier for example, does not appear to react in the same way, possibly because its gradient is not as steep.

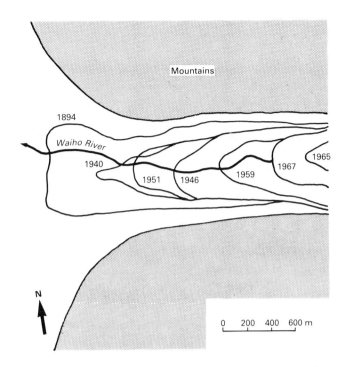

Fig. **6.21** The Franz Joseph Glacier: positions of the snout 1894–1967

Period	I Number of years	II Was the glacier snout advancing or retreating?	III Distance covered in metres	IV Average annual rate of change (III ÷ I)
1894–1940				
1940–1946				
1946–1951				
1951–1959				
1959–1965				
1965–1967				

Table **6.2** Details of changes in the position of the snout of the Franz Joseph Glacier (for use with exercise 15)

16 a) Can you suggest why heavy winter precipitation should lead to an eventual advance of the snout of the Franz Joseph?
b) Following the 1967 advance a moraine 1 m high was left across the valley. What type of moraine would this have been? Look at the sketch map once again and suggest why this moraine may not remain for very long.
c) Look at your atlas and try to determine the approximate mean annual precipitation of the region. With much of this in the form of rainfall, can you suggest why this amount might affect the permanence of any moraines?
d) Where on Fig. 6.21 would you expect to find boulder clay deposits? Justify your answer.

of stones is needed to draw valid conclusions. In collecting the stones, it is important that the position of stones being studied is not changed. The findings are best grouped and represented on polar graph paper (Fig. 6.23); this will enable a 'rose diagram' to be drawn, which will provide a good visual representation of ice movement.

Orientation of glacial stones' long axes in degrees from magnetic North		Number of stones
A	B	
0	180	25
20	200	55
40	220	35
60	240	3
80	260	4
100	280	19
120	300	17
140	320	8
160	340	9

Table **6.3** Details of a sample of 175 glacial stones from boulder clay deposits in North Norfolk

Table 6.3 shows the results of a stone sample in a North Norfolk boulder clay deposit. You will see from the table that the results have been grouped; thus all stones with an orientation between 350° and 9° from magnetic north are shown in the 0° sector, those between 10° and 29° are in the 20° sector, and so on.

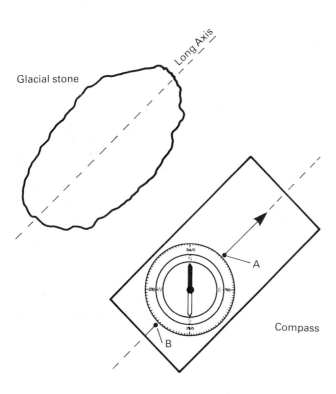

– Compass is aligned with the long axis of the glacial stone
– Central dial is set to magnetic north
– Orientation is determined by reading off the values at A and B on the compass

Fig. **6.22** Method for determining the orientation of glacial stones in boulder clay

17 Study Table 6.3.
a) Using polar graph paper, prepare a diagram similar to that shown in Fig. 6.23 and on it plot the results of the stone sample, using dots. For each group there must be two plots (column A and column B), representing each end of the stones' long axes. The plots for 0° and 180° have already been made. When complete, join up the dots and shade in the enclosed zone to produce a rose diagram.
b) What conclusions can you draw about the probable direction of ice movement at the time the deposits were laid?

Stones found in boulder clay are often aligned in the direction of ice flow. Fig. 6.22 shows how the orientation of a glacial stone can be determined. Care must be taken in selecting boulder clay which has been undisturbed by farmers and other land users. A sample

The stones embedded in boulder clay deposits will frequently be of a completely different rock type to the bedrock on which the deposits are laid. This is because the ice will have transported them considerable distances from their point of origin. Such stones are

Glacial Processes and Associated Landforms

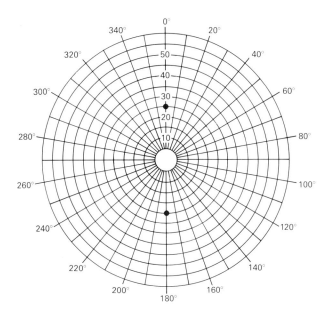

Note: Each successive circle is equivalent to 5 stones.

Fig. **6.23** Polar graph (for use with exercise 17)

Fig. **6.24** A glacial erratic near Ingleborough, North Yorkshire

known as **erratics** and by comparing their present location with their probable point of origin, a good picture of ice movement can be established.

Fig. 6.24 shows a fine example of an erratic, with the dark-coloured block of Silurian Grit contrasting with the lighter-coloured Carboniferous Limestone. Of course, erratics are not generally so large, nor is the bedrock usually exposed to make recognition so easy. In this case the limestone has slowly weathered away since the erratic was dumped, so that the plateau is now 50 cm lower than before. The effect has been to leave this particular erratic as a **perched block**.

Fig. 6.25 shows the locations of a number of glacial erratics found on a stretch of the North Yorkshire coast. The coastal strip here is covered with boulder clay and this has meant that the limit of ice advance during the Devensian (refer back to p. 135) can be established with some accuracy.

18 Locate this stretch of coastline in your atlas.
a) Locate the four main source regions of the erratics. On a sketch map showing Britain and Scandinavia mark the probable directions of ice flow.
b) Using the scale shown in your atlas, measure the distances involved between the source regions and points of dumping.
c) What conclusions can you draw about the nature of ice flow in the Devensian?

Occasionally boulder clay is dumped in the form of **drumlins**. These are large, elongated mounds of moraine, several hundred metres long, which have been streamlined by ice flow. Their origin is not entirely understood, but they are found on lower-lying land, often where a glacier may have fanned out from the highlands (forming a **piedmont glacier**). They occur in swarms, producing a distinctive 'basket of eggs' landscape, as shown in Fig. 6.26 (a).

Drumlin long profiles are usually asymmetrical, with the steeper slope facing the advancing ice. Cross-profiles are much more symmetrical. The contour diagram of Risebrigg Hill near Barnoldswick in North East Lancashire (located on Fig. 6.27) will allow you to examine this.

19 Study Fig. 6.26 (b).
a) Draw the long and cross-profiles of the drumlin in your exercise book on copies of sections 2 and 3.
b) Measure the length and height of the drumlin in metres.
c) From which direction was the ice advancing? Justify your answer.

Fig. 6.27 is a map showing the contour outline of drumlins in the Barnoldswick area. You will see a considerable variation in size and shape, often quite different from the classical drumlin form represented by

Glacial Processes and Associated Landforms

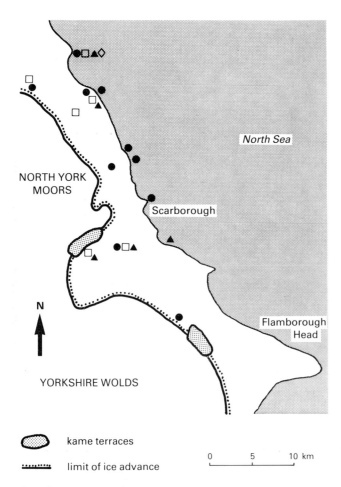

kame terraces

limit of ice advance

●□▲◇ locations of glacial erratics

source regions of erratics:
● Cumbrian Mountains, particularly Shap
□ Southern Scotland and the Cheviot Hills
▲ Scandinavia particularly Norway (Oslo district)
◇ Grampian Mountains

Fig. **6.25** Glacial erratics in North Yorkshire

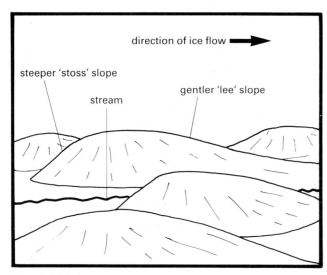

Fig. **6.26 (a)** A swarm of drumlins produces a characteristic 'basket of eggs' landscape

Risebrigg Hill. Since drumlins were shaped by ice flow, the orientation of their long axes provides a good indication of ice movement. The long axes of two drumlins in the west of the map area have been shown by means of a thin line.

20 Study Fig. 6.27 and locate the region in your atlas.
a) Determine the long axes of the remaining drumlins.
b) What conclusions can you draw about the direction of ice movement?
c) Using information on the map and in your atlas, particularly the relief of the region, suggest reasons for any differences in ice movement you detect.

You have now considered a number of ways of detecting past ice movements from landscape evidence: cirque alignment in Snowdonia, stone orientation in North Norfolk boulder clay, source regions of erratics in North Yorkshire and drumlins near Barnoldswick.

21 Re-examine your results for these four exercises.
a) Using Fig. 6.4 examine the nature of ice flow during the Devensian in relation to the four regions.
b) Do your results agree with the directions shown on the map? Would you wish to modify or add directional arrows in the light of your results?

Fluvio-glacial Deposition

Ice wastage at the glacier's snout produces a great discharge of water in the form of meltwater streams. These may actually originate and flow within, below, or on top of the ice itself as **englacial**, **subglacial** or **supraglacial streams**. However, in front of the snout they tend to flow in a very disorganised, **braided** fashion, constantly uniting and subdividing, across a broad plain known as an **outwash plain**.

The importance of these meltwater streams is that they can transport and redeposit glacial debris and thus create another set of landscape features. These deposits are known as **fluvio-glacial deposits**. They differ from

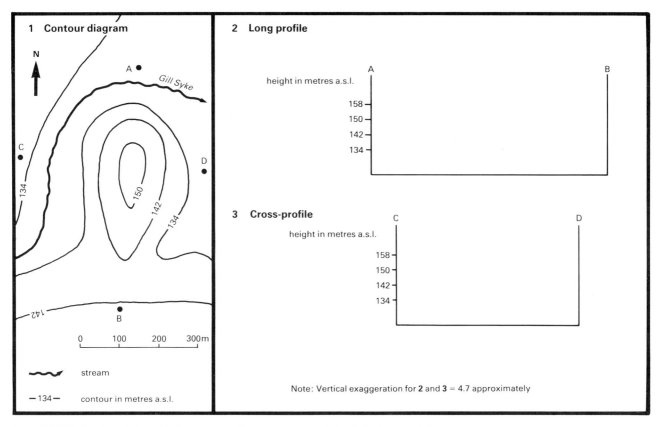

Fig. **6.26 (b)** Risebrigg Hill – a drumlin near Barnoldswick, Lancashire

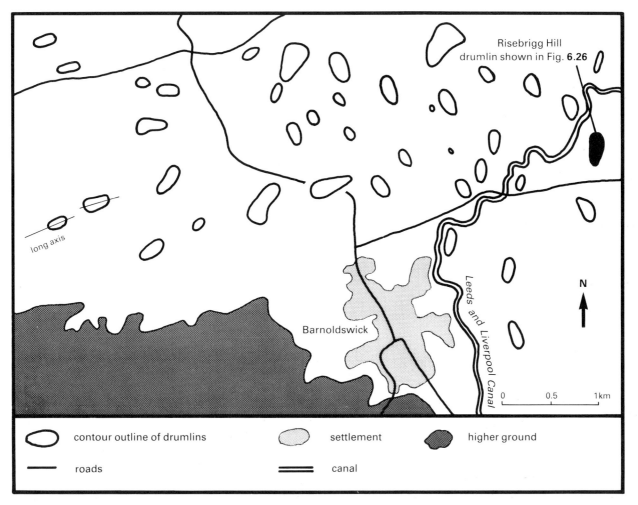

Fig. **6.27** Drumlins in the Barnoldswick area

the morainic deposits discussed above by being **stratified**, showing some degree of layering and sorting; moraines are unstratified, having simply been dumped by the ice. Since running water is involved in this process, fluvio-glacial deposits tend to be much more rounded and smoothed, consisting mostly of sands and gravels; morainic stones by contrast are rough-edged and angular.

> **22** Fig. 6.28 shows the North Iliamna Glacier in Alaska.
> a) In your exercise book draw a sketch of this photograph.
> b) On the sketch identify the following features: i) outwash plain, ii) braided meltwater streams, iii) glacier snout.
> c) How would you account for the dark, curved, ridge deposits in the centre of the photograph? What do they suggest about the recent history of the glacier? Notice how they have been breached by the streams.

Most fluvio-glacial deposition takes place in front of the glacier, producing few distinctive features apart from the outwash plain itself. The collective term for such deposits is **outwash deposits**.

Fluvio-glacial deposition in contact with the ice is more recognisable. Mounds of fluvio-glacial deposits are known as **kames**; they have no uniform shape. When deposition takes place by supraglacial streams or lakes against the valley side, a **kame terrace** may result, as shown in Fig. 6.29. Sediment gradually builds up in the ice-marginal lake to form a terrace which remains when the ice retreats. The locations of some example kame terraces are shown on Fig. 6.25. Notice their location in relation to the ice limit.

Long winding ridges of sand and gravel fluvio-glacial deposits are known as **eskers**. They were probably formed by meltwater streams flowing in ice tunnels beneath the glacier. Fig. 6.30 shows one of several eskers in the Trim region of Ireland which are featured in Fig. 6.31. The winding character of the ridge is highlighted by the contrasts in vegetation and the course of the road which hugs its lower slopes, keeping off the wetter plain. Since the meltwater streams would have flowed in the same direction as the glacier, the alignment of eskers reflects the direction of ice movement.

Fig. **6.28** The North Iliamna Glacier, Alaska

Glacial Processes and Associated Landforms

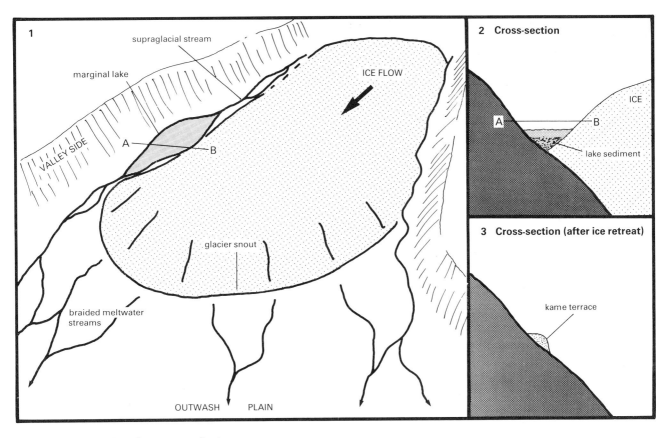

Fig. 6.29 The development of a kame terrace

23 Fig. 6.31 shows a map of the eskers in the Trim region of Ireland.
a) Locate the region in your atlas (it is just to the north-west of Dublin). Using information on the map and in your atlas (e.g. relief) can you suggest which direction the ice was flowing at the time of esker formation?
b) Why do you think the eskers are not continuous?
c) Measure the length of the eskers (including the breaks) and calculate the mean length for the system shown.
d) The Trim esker (AB) is 4.6 m high at A and 15 m high at B. In the light of your conclusions about ice movement, attempt an explanation for the difference in height.

Fig. 6.30 An esker system in Trim, Co. Meath, Ireland

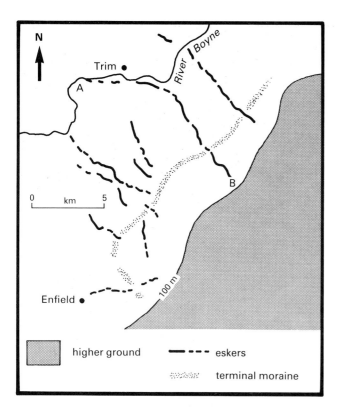

Fig. **6.31** Eskers in the Trim region of Ireland

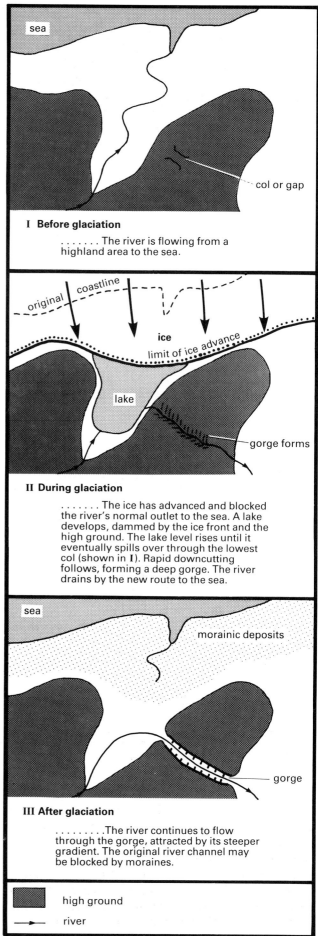

Fig. **6.32** Glacial diversion of drainage

Glacial Diversion of Drainage

Glaciers may have significant *indirect* influences on the development of landscape features. Perhaps the most important of these result from glacial diversion of drainage, when a river's normal course to the sea is completely altered by glacial interference. Fig. 6.32 shows a typical sequence of events leading to glacial diversion. Such diversions are relatively common in Britain which was on the margins of the major glacial expansions.

One of the best known examples of glacial diversion of drainage is that of the River Severn which rises in Mid Wales, flows northwards and then doubles back on itself to enter the Bristol Channel to the south. This is shown in Fig. 6.33.

Some geomorphologists believe that the Goring Gap on map extract 1 is also an overflow gorge from a so-called 'Lake Oxford'. From a study of the map extract, you should be able to determine where, in relation to the chalk scarp, this lake would have been.

Glacial Processes and Associated Landforms

24 Study Figs. 6.32 and 6.33 carefully.
a) In your exercise book draw your own sequence of diagrams to show how the Severn may have been diverted.
b) Mark on where you consider the lake to have existed and name it 'Lake Lapworth'. Can you suggest what evidence you would look for *in the field* to show the extent of this lake and the limits of the ice front?
c) Mark also where you would expect to find an overflow gorge and suggest where the Severn may have originally entered the sea.

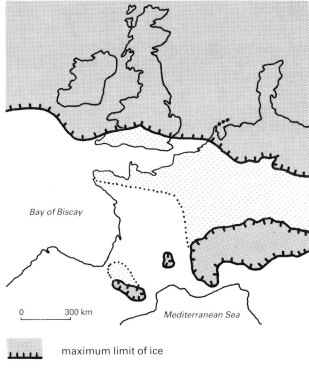

Fig. **6.34** Maximum limits of ice and permafrost in Western Europe during the Pleistocene

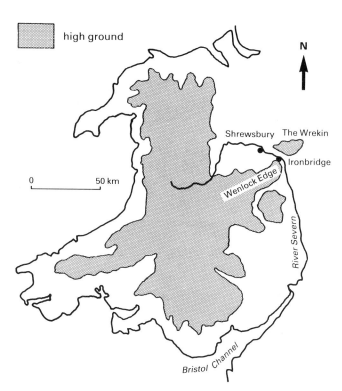

Fig. **6.33** The course of the River Severn

south (see Fig. 6.3). So whilst southern England never experienced glaciation, it did come under the influence of periglacial conditions.

A major feature of a periglacial environment is the extent of frost action. Frost may penetrate deep into the ground which may become permanently frozen, a condition known as **permafrost**. Only the surface layer, the **active layer**, will actually thaw out during the brief summer. Furthermore, frost action will be responsible for freeze-thaw weathering, producing much rock debris.

Periglaciation

Regions immediately around ice sheets and ice caps are referred to as **periglacial regions** (literally meaning 'around the ice') or **tundra**. Here temperatures are for the most part below freezing point, but a brief summer period allows thawing to occur. Today the periglacial regions are confined to the extreme northern latitudes of America and Eurasia, as you will see if you refer back to Chapter 4 (p. 92), but during the Devensian and at other times in the Pleistocene they were much further

25 Fig. 6.34 shows the maximum limit of permafrost development during the Pleistocene in Western Europe.
a) Compare this with the present distribution of tundra in Europe. You will find this by referring back to Fig. 4.19 (p. 79). All points between the present tundra regions and those in Fig. 6.34 should therefore contain some evidence of periglaciation.
b) Can you suggest why ice caps developed in the south of Western Europe?

The presence of permafrost may have important consequences for the development of landforms. If the ground is sloping, then during the summer the active layers, lubricated by meltwater, can easily flow over the frozen layers below. This process is known as **solifluction** and will lead eventually to the build-up of deposits downslope. These deposits consist of angular rock fragments produced by freeze-thaw weathering and are called **head deposits**. The main difference between head deposits and boulder clay described earlier in the chapter, is that head contains virtually no erratics, being closely related to the local rock type. This is because solifluction movements only occur over short distances.

Since water cannot pass through ice, permeable rocks experiencing permafrost conditions become impermeable. Rather than seeping through the rock, rain will begin draining off the surface. When this happens, rivers form and valleys develop in the normal way (see Chapter 5). With the warming of the climate, the rock will become permeable again, leaving the valley that developed during the periglacial phase unoccupied by a river; such a valley is known as a **dry valley**, introduced in Chapter 2 (p. 35).

> **26 a)** Refer back to Chapter 2 and compare the explanation of dry valley formation given there with the periglacial explanation outlined above.
> **b)** Study the geological map on p. 26 in conjunction with map extract 1. Describe the distribution of the head deposits in relation to the relief of the region. What rock type would you expect these deposits to consist mainly of?

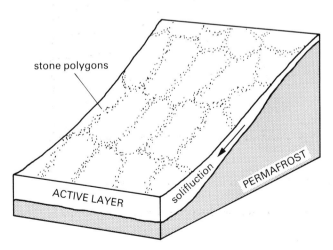

Fig. 6.35 Patterned ground in a periglacial region

The development of permafrost often causes the ground to heave, since the water within the ground expands on freezing. This heaving may allow the surface rock fragments to become arranged in distinctive polygonal shapes, creating **patterned ground** (as shown in Fig. 6.35). Stone polygons are common in today's tundra regions such as northern Alaska. They are small features, only a few metres in diameter.

Periglacial regions are bleak and contain little vegetation. Winds can be strong and may pick up the finer deposits and redistribute them beyond the permafrost zone. Such wind-borne deposits are known as **loess** (or **limon** in France) and they are quite common in central Europe, though virtually non-existent in Britain.

The Complexity of Glaciation

Glaciated landscapes are invariably more complex than one might first imagine and erosional and depositional features may often exist together. Fig. 6.36 shows Cwm

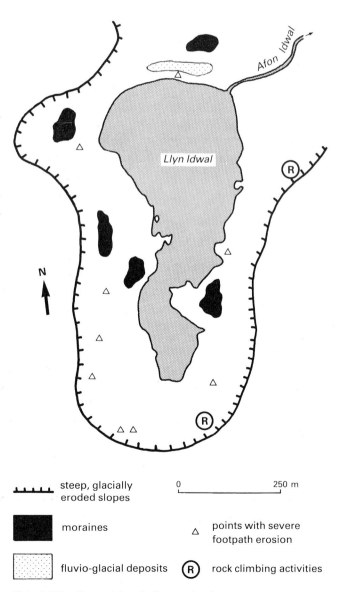

Fig. 6.36 Cwm Idwal, Snowdonia

Glacial Processes and Associated Landforms

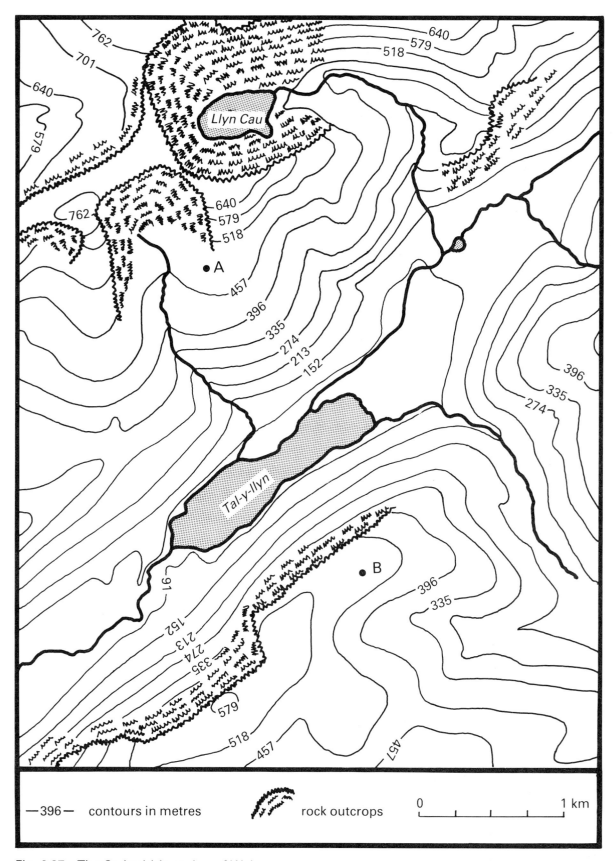

Fig. **6.37** The Cader Idris region of Wales

Idwal in Snowdonia, a fine example of a cirque with a tarn. Whilst the cirque exhibits the steep abraded slopes associated with glacial erosion, it also contains a number of hummocky morainic deposits, probably representing examples of lateral and recessional moraines. The fluvio-glacial deposits shown are remnants of a kame terrace. Solifluction has certainly been active in the cirque and freeze-thaw weathering continues to be responsible for the scree slopes.

The cirque is an important recreational area and has come under a great deal of recent pressure from groups of people (rock climbers, etc.). This pressure has led to severe footpath erosion in several places confirming that man, like ice, is a geomorphological agent.

> **27 a)** For the deposits shown on Fig. 6.36, explain what characteristics you would look for to determine their likely mode of origin.
> **b)** In addition to rock climbers, name other groups of people who probably exert pressure on the cirque.
> **c)** What geomorphological agents other than man are likely to be operating in the cirque today?
> **d)** Describe the effects of these agents on landform development.

Glaciation and Man

Glaciated landscapes are of great significance to mankind. Glaciated uplands are scenically very attractive and in Britain they represent major areas for recreation and tourism, a vital source of income in regions with limited economic potential. It is no coincidence that the two finest glaciated upland regions in England and Wales, namely the Lake District and Snowdonia, have been designated as **National Parks**, areas which enjoy great protection from urban development. Glaciated U shaped valleys can be easily dammed and, along with natural ribbon lakes, they provide ideal conditions for hydro-electric development.

Glaciated lowlands are equally important. Many sand and gravel outwash deposits have been quarried and used extensively in the construction industry. Boulder clays and loess are often extremely fertile and provide excellent conditions for farming. We have seen above how eskers may, in certain conditions, make good lines of communication, whilst the city of York stands on a morainic ridge overlooking the lower-lying vale.

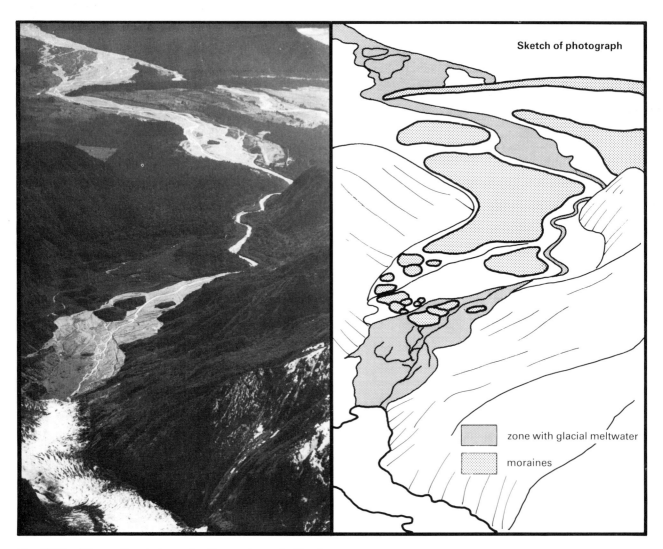

Fig. **6.38** The valley below the Franz Joseph Glacier

Finally, glacial overflow channels may be of considerable importance in controlling routeways; the influence of the Goring Gap on communications has already been noted.

Workshop

1 Study Fig. 6.37 which shows a number of glacial features in the Cader Idris region of Wales. Re-read the section in this chapter on glacial erosion.

a) Draw an annotated sketch of the map, locating and identifying as many glacial features as you can.

b) Draw a cross-profile across the valley containing Tal-y-llyn from point A to B. Choose a suitable vertical scale for your profile and calculate the vertical exaggeration. Calculate the average gradients of the valley sides using the information on the map and the method shown on page 116. Comment on your findings.

2 Fig. 6.38 shows a photograph looking down-valley from the Franz Joseph Glacier in New Zealand which we studied earlier in the chapter. On a copy of the accompanying sketch diagram, identify the following features from the photograph: a) valley glacier, b) U shaped valley, c) arête, d) outwash plain, e) recessional moraine, f) lateral moraine, g) terminal moraine. Briefly describe how each feature developed.

3 Fig. 6.39 shows an outline map of a glaciated lowland and mountain front with a valley glacier, during a period of climatic warming; the limit of ice advance is marked. On a large copy of this map indicate, using suitable symbols, where you might expect to find the following deposits: a) boulder clay, b) erratics, c) terminal and recessional moraines, d) a swarm of drumlins, e) eskers, f) kame terraces, g) outwash deposits, h) head. (N.B. you should realise that it is very unlikely that in reality all these deposits will be found in such close proximity to one another.)

4 Try to find any recently felled tree in your local region and make a study of its ring system. If past local weather records are available you can try relating size of tree ring to disturbances in weather pattern. You ought, for example, to be able to spot the 1976 ring when the summer was exceptionally hot and dry.

5 Obtain a winter sports holiday brochure and make a study of the ski resorts it is advertising. Look particularly at photographs and diagrams of the ski-runs. Write a description of any glacial features you observe and comment on the link between the landscape and its development by man.

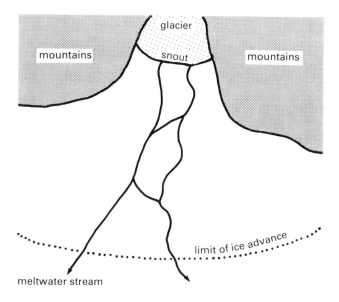

Fig. **6.39** Outline map for showing lowland glacial features (see Workshop exercise 3)

Coastal Processes and Associated Landforms 7

The enormous variety of coastal scenery in the British Isles is a reflection of the many geomorphological processes and factors involved. However, it is the action of the sea, and waves in particular, coupled with variations in rock type, which are the main controls.

Wave Action

The character of the waves reaching the coast determines whether erosion or deposition will take place. Important features of waves are **height**, the vertical difference between crest and trough, and **wavelength**, the horizontal distance between successive crests (Fig. 7.1). Waves simply represent the circular or orbital movement of water particles without any horizontal movement of the water itself (in the same way as a sound wave travels through the air without displacing it) until the wave breaks. This orbital movement declines with depth, so that below a depth equivalent to one wavelength, orbital movement is negligible. Wave energy is therefore concentrated in the surface layers of the sea; big waves contain more energy than small ones and can therefore do more geomorphological work.

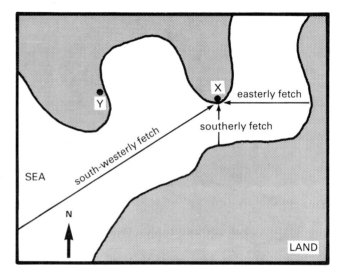

Fig. 7.2 Variations in fetch for a location

Most waves develop from wind blowing across the water's surface. Strong winds produce big waves, though there are other controls. Waves are not produced spontaneously but take some while to form; the bigger the wave, the greater the length of time

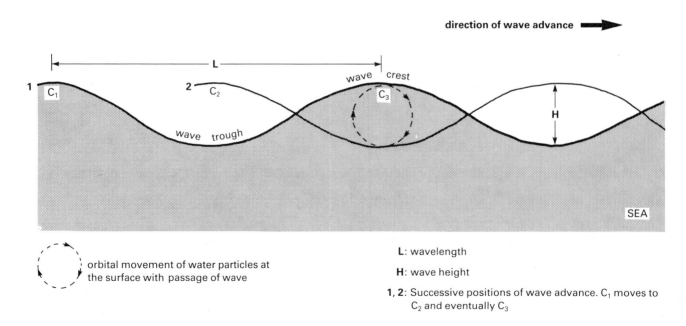

L: wavelength
H: wave height
1, 2: Successive positions of wave advance. C_1 moves to C_2 and eventually C_3

Fig. 7.1 Wave characteristics and measurements

Coastal Processes and Associated Landforms

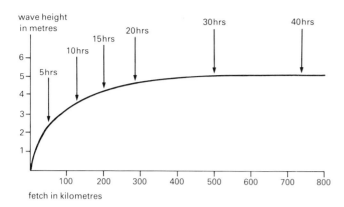

Fig. **7.3** The relationship between wave height, length of fetch and time for a wind blowing constantly at 50 km per hour

required. If the wind changes direction during this period, the process of wave formation may be upset. Thus **time** and wind **constancy** exert important controls over wave development. The expanse of water over which a wind may blow without interruption is also vital. This is known as the **fetch** and no matter how powerful the wind, if the fetch is small, wave height will always be restricted. Points on the coast will have a range of fetches. In Fig. 7.2 point X will never experience great wave attack with easterly and southerly winds blowing. However, the great south-westerly fetch will allow winds from that direction to build up big waves which will be capable of great erosion. The diagram shows how some locations (e.g. Y) will always be more sheltered. The waves related to the greatest fetch are known as **dominant waves**; **prevailing waves** are those occurring from the most common direction.

1 Fig. 7.3 shows how the above factors combine to influence wave height when a 50 km/hr wind is blowing. Study the graph carefully and answer the following.
a) What is the maximum wave height possible without restrictions of time or fetch?
b) With an indefinitely constant wind, what will be the wave height with a fetch of 200 km?
c) With an unlimited fetch, what is the maximum wave height possible with the wind blowing for only 10 hours?
d) What is the prevailing wind direction for the British Isles and in which direction is the greatest fetch?
e) Name some locations in Britain you know of that are exposed to great wave attack and briefly explain why.

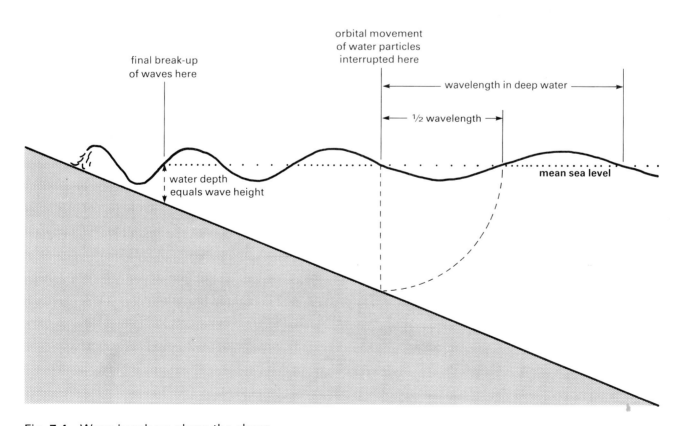

Fig. **7.4** Wave break-up along the shore

When a wave approaching the coast reaches a mean depth of water approximately equal to half its wavelength, the orbital movement of all water particles in the wave will be interrupted. This causes the rate of wave advance to slow down, the height of the wave increases and eventually it will start to break up when the mean depth of water is roughly equal to wave height. Fig. 7.4 shows this. It is during the break-up of the wave that energy is released.

As a general rule, high energy levels result in erosion whilst low energy levels lead to deposition. But the release of wave energy along the coast is rarely constant, largely because of the irregular nature of the coastline itself. Fig. 7.5 shows what happens to a wave (W) when it advances on a coastline of alternating headlands and bays, assuming all points on the coast have a similar offshore gradient. Since the sections of the wave opposite the headland will reach shallow water first, they will be the first to slow down and break up, with the break-up of those sections advancing into the bay occurring later. The result is a considerable distortion of the wave crest (W_4), known as **wave refraction**.

If we assume that initial energy levels along the wave are evenly spread, we can examine the effects of this wave refraction on the final distribution of energy along the shoreline. This can be done by formulating a simple hypothesis.

Hypothesis: *that more wave energy will be concentrated on headlands than in bays.*

On Fig. 7.5 you will notice that a pecked line has been drawn through successive positions of the wave crest, from an original position on wave W, marked with a dot.

2 a) In your exercise book draw, as carefully as possible, a copy of Fig. 7.5.
b) For all other dots on the wave W, draw in the pecked lines so that each cuts W_1, W_2, W_3 and W_4 at right angles. Mark dots where each of these pecked lines hits the shoreline.
c) Describe the pattern of wave energy hitting the shoreline. Where is energy concentrated and where is it dispersed?
d) Does this pattern lead you to accept or reject your hypothesis?
e) In view of the link between energy levels and the processes of erosion and deposition, does this exercise offer any explanation for the beach deposits in the bays?
f) How might these differences in the distribution of wave energy eventually affect the form of this indented coastline?

Wave refraction may also occur when waves advance *obliquely* to the general trend of the shoreline. Fig. 7.6 shows the typical extent of wave refraction around a headland. Notice that although the waves originate from a north-westerly direction, they hit the coast at point X from a near easterly direction. Therefore, although X is in a sheltered location it will not escape breaking waves.

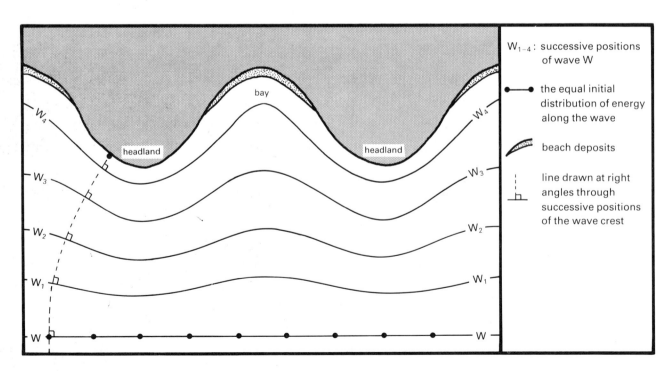

Fig. 7.5 Wave refraction on a headland/bay coastline

Coastal Processes and Associated Landforms

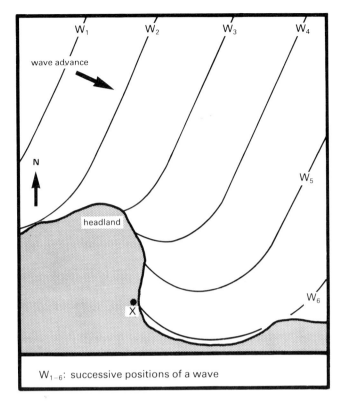

Fig. **7.6** Wave refraction around a headland

> **3 a)** Using the method employed in exercise 2, show how wave refraction may affect the distribution of wave energy along the shoreline shown in Fig. 7.6.
> **b)** What will be the effect of refraction on coastal processes along this particular shoreline?

When a wave breaks, the uprush of water on the shoreline is termed the **swash**, followed by the **backwash**, the return of the water downslope. Variations in the relative strengths of wave swash and backwash allow two main types of wave to be recognised, the **spilling wave** and the **plunging wave**, shown in Fig. 7.7. In a spilling wave, wave height is generally low and since the wave breaks at a gentle angle, swash is more powerful than backwash. In plunging waves, however, wave height is greater and so the wave breaks at a steeper angle, producing a more powerful backwash than swash. These differences mean

Fig. **7.8** Breaking waves

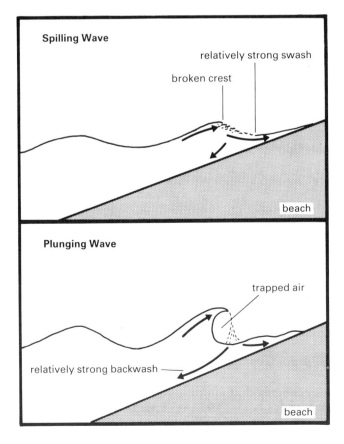

Fig. **7.7** Types of wave action

that spilling waves tend to be constructive, leading to a build-up of beach material onshore, whilst plunging waves are destructive and cause a removal of such material.

> **4** Identify the types of wave shown in Fig. 7.8 and the processes you think may be in operation.

Marine Erosion

The sea acts as an agent of erosion. The **hydraulic action** of the breaking waves can cause considerable erosion. The great mass of water crashing against an obstruction can easily loosen and dislodge rocks, as may the air which the wave traps under great pressure in joints and fissures. In addition, breaking waves can pick up eroded material and hurl it against the shoreline. This will lead to the **corrasion** (abrasion) of the coast as rock particles break off others. Eventually these rock fragments will be worn down themselves into smaller, finer sediments; this process is known as **attrition**. Finally, certain rocks such as limestone will always **corrode**, i.e. react chemically with the sea water.

We must now examine the effects of continued erosion on landform development along a shoreline. We will consider a sloping shoreline containing no pre-existing coastal features. Fig. 7.9 shows the typical sequence of events and as such can be described as a **model of coastal evolution**.

The two main features formed are the **wave-cut platform** (or **abrasion platform**) and the **cliff**. The wave-cut platform tends to be slightly concave in profile, irrespective of the type and structure of the rock. The wider the wave-cut platform, the more wave energy will be dispersed, leaving less for undercutting at the **cliff-foot notch**. This may often take the form of a cave, particularly where a point of weakness in the rock exists. The form of the cliff above will be related to a) the rate of undercutting by waves, b) the natural resistance and structure of the rock, and c) the degree of slumping, rainwash and weathering on the cliff face.

Fig. 7.10 shows two different views of the wave-cut platform in Clarach Bay, Mid Wales. This is located at 586842 on map extract 2. The wave-cut platform has been cut into steeply dipping, resistant sedimentary rocks. Notice in Fig. 7.10 A the slight slope of the platform from the cliff foot, where a small notch is visible in the centre, though some beach deposits exist to either side. In Fig. 7.10 B the seaward dip of the strata is apparent, as is the smooth, abraded surface of the exposed rock and the relative lack of loose pebbles and boulders.

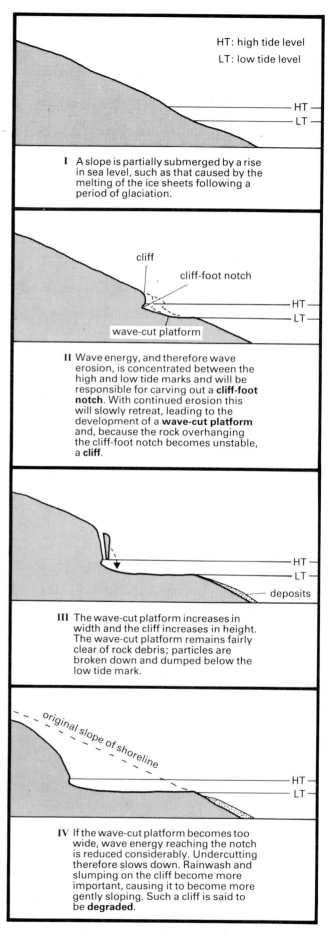

Fig. **7.9** A model of coastal evolution

Coastal Processes and Associated Landforms

Fig. **7.10** The wave-cut platform at Clarach Bay, Mid Wales

Fig. **7.11** Cliffs to the north of Clarach Bay

Fig. 7.11 shows two views of the cliffs a little further to the north of the bay. In Fig. 7.11 A the strata on the headland are dipping landwards (by contrast to those on the wave-cut platform) and this dip has clearly influenced the profile of the cliff. The abrupt change of slope or **bevel**, at the top of the cliff (at X), is caused by the presence of weak, glacial deposits overlying the sedimentaries; such deposits are much more prone to slumping. Fig. 7.11 B shows a cave. You can see this has been eroded at a fault-line in the rocks, marked not by a clean shear plane, but by a shatter zone which presents a major point of weakness to wave attack. By following the strata identified X and X_1 on the left of the cave, to Y and Y_1 on the right, the extent of the fault can be appreciated.

> 5   Study carefully Figs. 7.10 and 7.11.
> a)   Draw sketches of at least one photograph in each figure.
> b)   For each sketch, identify all the relevant features that have been discussed and annotate it accordingly.

Wave attack is therefore selective, with joints, bedding planes, as well as faults, representing the key points of weakness. We should now examine the features that may result from a continuous period of marine erosion on such weaknesses. Fig. 7.12 shows a typical sequence of landform development on a headland (i.e. a **model of headland evolution**), which produces in turn a **cave**, **arch**, **stack** and **stump**. Such features are unlikely to develop in weak rocks such as clays. Occasionally in the development of a cave, erosion may cause the ceiling of the headland to be breached up a vertical joint, forming a **blow-hole**. Vertical weaknesses on the side of the headland may be attacked and lead to the development of very narrow inlets or **geos**. Good examples of these features are

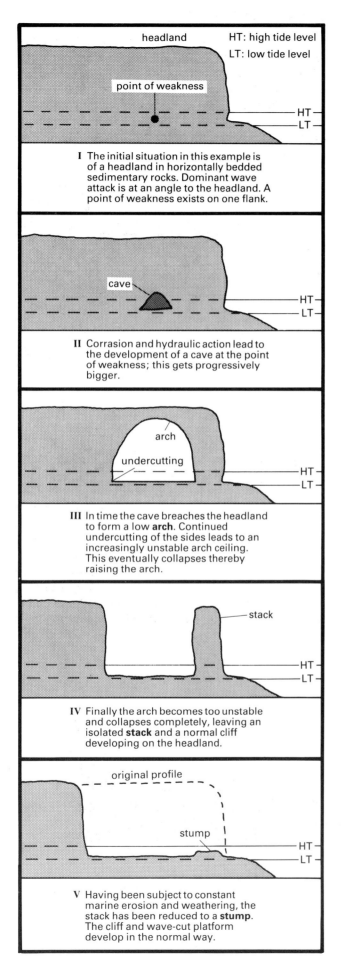

Fig. **7.12** A model of headland evolution

Fig. **7.13** The Green Bridge of Wales

Fig. **7.14** Elegug Stacks

Coastal Processes and Associated Landforms

found in the Carboniferous Limestone coastline of South Pembrokeshire (Dyfed); Figs. 7.13 and 7.14 represent some of the most dramatic.

> **6 a)** Study the photographs and write a detailed account of the evidence indicating that marine erosion has occurred.
> **b)** Consider the factors, particularly geology and structure, that appear to have influenced the form of the features shown.

We can now analyse the entire cross-profile of an erosional shoreline. Fig. 7.15 shows the cross-profile at Clarach Bay (considered above) drawn from measurements taken in the field. Note the feature marked Z can actually be seen in Fig. 7.10 A.

> **7** Study Fig. 7.15 and answer the following.
> **a)** What are the features marked S and Z?
> **b)** Using the scales provided measure i) the height of the cliff, ii) the width of the wave-cut platform, and iii) the heights of S and Z above the wave-cut platform.
> **c)** What is the angle of the cliff face? Compare your answer with Fig. 7.11 A.
> **d)** Using Fig. 7.10 A, explain why the landward side of feature Z is steeper than the seaward side.
> **e)** Using Fig. 7.9 try to establish where this shoreline fits into the model of coastal evolution.

Erosional Coastlines

Fig. 7.16 shows part of the Gower Peninsula, another area in Wales composed largely of Carboniferous Limestone, like South Pembrokeshire. Here a number of structural elements have contributed to the main coastal features and the shape of the coastline itself. You may have to refer back to Chapter 1 to remind yourself what some of the structural elements are.

> **8** Study Fig. 7.16 and answer the following.
> **a)** Describe and explain the position of the wave-cut platforms and cliffs in relation to the direction of the dominant waves.
> **b)** What effect have i) the anticline, and ii) the two synclines had on the shape of the coastline? Can you suggest reasons why?
> **c)** Describe and explain what has happened to the faults exposed on the coastline.
> **d)** Give an explanatory account of the factors influencing the location of the beach deposits.

Given the same level of wave attack, rocks of different resistance will erode at different rates. Thus when a number of contrasting rock types are exposed close to one another, the effect is to produce quite an irregular coastline.

Fig. 7.17 features part of the Dorset coastline, where the rocks are orientated roughly parallel to the shoreline. At one time there would have been continuous bands of all three rock types along the coast, but marine erosion has been progressively removing them. Where the outer bands of rock remain, it is

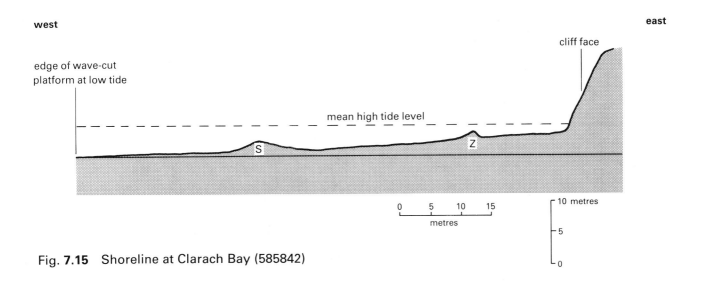

Fig. **7.15** Shoreline at Clarach Bay (585842)

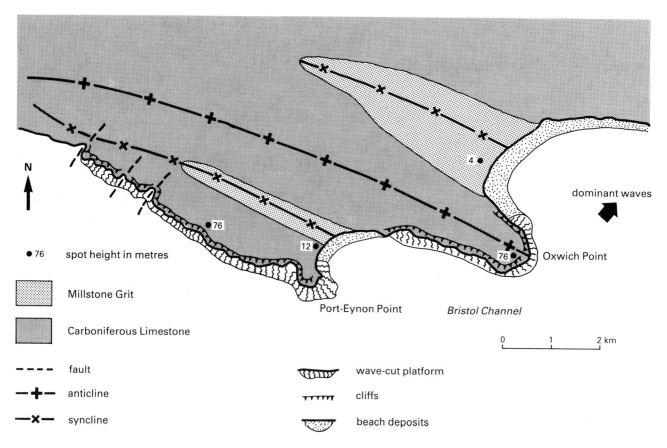

Fig. 7.16 Part of the Gower Peninsula coastline, South Wales

possible to examine the effects of **differential erosion**. The oldest rocks here are the Purbeck Beds, very resistant sedimentaries which dip steeply towards the land. Behind them lie the much weaker Wealden Beds which consist mainly of the easily eroded Wealden Clay. Finally, to the north is found the youngest and most extensive rock, chalk, which in places rises to over 150 m.

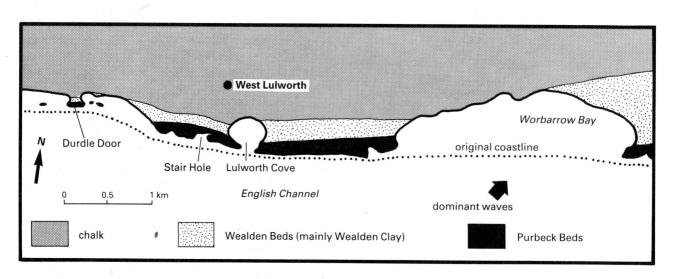

Fig. 7.17 The geology of part of the Dorset coastline

Coastal Processes and Associated Landforms

> **9 a)** What are the characteristics of rocks in general which influence their resistance to erosion?
> **b)** What is the evidence on Fig. 7.17 to suggest that the Purbeck Beds are more resistant than the Wealden Beds?
> **c)** Describe in your own words the marine processes taking place at Stair Hole and Lulworth Cove.
> **d)** What sort of coastal landforms would you expect to find in the Purbeck Beds? Why?
> **e)** Draw two maps to show what the coastline might look like i) in the near future, and ii) in the distant future. Explain the coastlines you draw by annotating them fully.

You can check your answer to question (d) by reference to Fig. 1.11 (a).

The Holderness coastline in Eastern England provides an interesting study of coastal retreat over many centuries. The cliffs here are composed mainly of glacial deposits and annual retreat in parts of southern Holderness *averaged* 2.75 metres for the period 1852–1952, one of the highest figures in the world. Naturally, substantial variations from this average are likely to occur, so that between 1951 and 1966 annual figures ranged from 0.15 to 10.3 m.

Losses of over ten metres of cliff in one day are not unheard of. They generally occur during **surge tide** conditions when, in the North Sea, spring tides (i.e. highest of the high tides) coincide with strong northerly winds, whipping up the waters and causing them to funnel into the narrower parts of the southern North Sea. Surge tides may be nearly three metres above normal spring tides. Probably the most famous surge of recent decades occurred on the night of 31 January/ 1 February 1953 and led to widespread erosion and flooding along the entire coast of Eastern England.

Old maps and records allow us to build up a picture of the likely position of the coastline nearly 2000 years ago, in Roman times. This is shown for Holderness in Fig. 7.18, along with the positions of former settlements which have been washed away during coastal retreat.

> **10** Study Fig. 7.18 and locate the stretch of coastline in your atlas.
> **a)** Briefly explain why glacial deposits should be relatively easy to erode (refer back to Chapter 6).
> **b)** From which direction will dominant waves advance?
> **c)** Try to calculate the area of land that has been lost to the sea during the last 2000 years.
> **d)** What was the mean annual loss of land?

Marine Deposition

When wave energy becomes too low for erosion or transportation, deposition of the marine sediment load will occur. Such deposition may be underwater, as we saw in Fig. 7.9, but often it is on the shoreline itself, taking the form of a beach. There are, however, some locations, e.g. headlands, where deposition is rare.

The form of beaches and the character of their deposits vary considerably. The main influencing factors are the nature and composition of the bedrock from which the deposits originate and the actual location of the beach. Since resistant rocks are not easily eroded, they tend to form shingle beaches, whilst when very weak rocks break down, the particles produced may be so small that little beach material can be provided. One of the problems on the Holderness coast is that only 10% of the cliff material contains particles large enough for the beach there. This means that a loss of 1 tonne of beach by destructive wave action requires the collapse of 10 tonnes of cliff to provide replacement material.

The coastline of Cornwall is quite different from that of Holderness. Granite is the major bedrock, producing a very rugged cliff-line. However, when granite is finally broken down, many small sand grains are produced, ideal for sheltered beaches. Fig. 7.19 is a tourist map of the extreme west of Cornwall on which are marked the

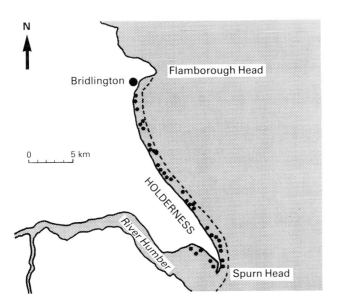

----- likely position of coast in Roman times
••• former settlements

Fig. **7.18** Coastal retreat in Eastern England

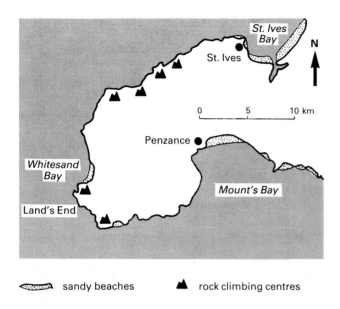

sandy beaches ▲ rock climbing centres

Fig. 7.19 The sandy beaches of West Cornwall

11 a) Suggest why granite should be generally resistant to wave attack. What is the mineral in granite from which the sand of Cornish beaches is derived? (Refer back to Chapter 1, p. 12.)
b) Why are the beaches in Cornwall often very dangerous for sea bathing? (Refer to an atlas showing Cornwall's situation in relation to the Atlantic Ocean.)
c) Why are there no sandy beaches on the stretch of coast between Whitesand Bay and St. Ives Bay?
d) Can you suggest any other rock types which might produce good material for sandy beaches?

sandy beaches favoured by holidaymakers (those with shingle have not been marked). These beaches have developed in locations where constructive wave action has operated over long periods and where shelter is available from destructive wave action. Since they are found in inlets, they are commonly known as **bay-head beaches**.

Beaches composed of pebbles frequently take the form of elongated **shingle ridges**, sometimes referred to as **storm beaches**. Here breaking waves have progressively piled the pebbles up on one another, so that the ridges become semi-permanent, rising to a height just above the highest spring tide. Storm waves have certainly been important in this process, since their

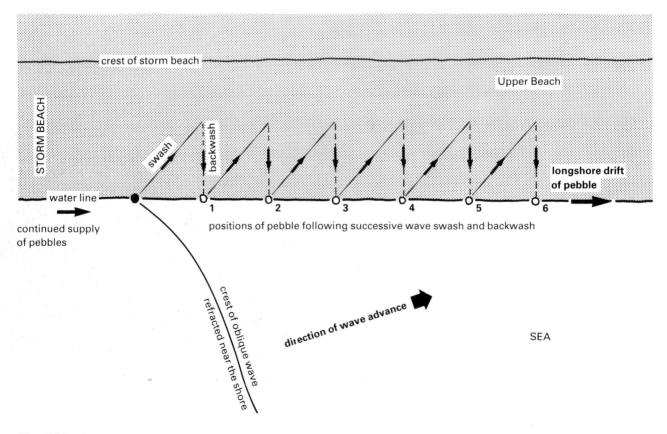

Fig. 7.20 Longshore drift

Coastal Processes and Associated Landforms

energy is great enough to hurl pebbles well up the beach. The source of the pebbles for storm beaches will probably have been along the coastline; oblique wave action would have been responsible for transporting the pebbles along the coast, a process known as **longshore drift**.

Fig. 7.20 shows how longshore drift may occur. Even though oblique waves will experience some 'straightening out' from wave refraction close to shore, their swash will still be diagonally up the beach. Naturally, if the energy is great enough, pebbles will be transported by this water movement. Backwash will take the line of maximum gradient of the beach, returning the pebbles to the water line. In this way, given the continued supply of pebbles, the storm beach will grow longer. The diagram shows that swash need not extend right up the beach. For much of the time longshore drift activity will be confined to the lower beach, leaving the upper beach inactive for long periods.

Let us consider the storm beach in greater detail. Fig. 7.21 shows a storm beach near Borth, West Wales. It can be located on map extract 2, running from Upper Borth in the south to Twyni Bâch in the north. The beach provides the basis for a number of useful studies. The picture was taken at 604930 looking north and it shows clearly the elongated nature of the beach. The fact that longshore drift has occurred is confirmed by the enormous build-up of pebbles on one side of the groyne shown in Fig. 7.22. The groynes are marked on the map extract.

Table 7.1 gives the data obtained from two cross-profile surveys of the Borth storm beach, undertaken 1 km apart at 604928 and 603938.

Fig. **7.21** The storm beach near Borth, West Wales

12 a) From Fig. 7.22 and map extract 2, work out the direction of longshore drift. Justify your answer.
b) Consider the likely source of supply of the pebbles on this storm beach.

The slope across a storm beach is rarely even. To establish the nature of any irregularities, a simple surveying method can be employed. The only equipment required is a) a clinometer, an instrument for measuring slope angles, which can be home-made (the one used in this exercise will need to be of a slightly different design to that used in Chapter 2, since small slope sections, rather than large ones, are being examined); b) a tape-measure; c) a flat-sided, rigid piece of wood or pole; and d) some distinctive markers (e.g. painted or coloured pebbles). The details of the surveying method are given in Fig. 7.23. A table of results can be prepared from such a survey and a profile can then be drawn.

Fig. **7.22** A groyne on the Borth storm beach

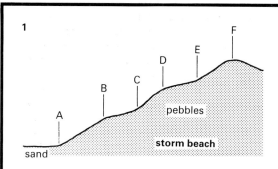

- Identify the points on the storm beach where slope changes (marked A to F above).
- Mark these points carefully e.g. with a distinctively coloured pebble.

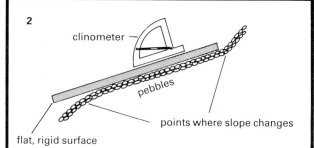

- On each slope section (A B, B C etc.) place a suitable flat, rigid surface e.g. a piece of wood. This is to even out minor irregularities.
- Measure the angle of the slope using the clinometer.
- Measure the distance between the two points using a tape measure.
- Record the measurements.

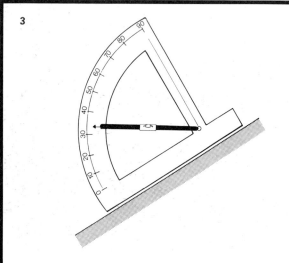

- Clinometers vary in sophistication. The one shown above is home-made from a large wooden protractor and a small spirit level. (Others can measure to a fraction of a degree.)
- Adjust the clinometer so that the bubble of the spirit level is centralised.
- Read off the angle on the scale.

Fig. **7.23** Surveying the cross-profile of a storm beach

13 Study the data in Table 7.1 carefully and acquire a protractor, ruler, sharp pencil and some metric graph paper.

a) The horizontal and vertical scale chosen for each profile should be 1 cm = 1 m (i.e. 1:100). Draw in the axes approximately 25 cm long by 6 cm high.

b) Start each profile at the origin of the axes. This will be point A. Then, for profile 1, you should start by measuring from point A an angle of 14° from the horizontal axis using the protractor. Draw in a feint line and measure 2.92 cm *along that line* from A (not horizontally). This will be point B.

c) From point B, measure an angle of 9.5° and draw in a line 1.53 cm long to point C.

d) Repeat this procedure for all the points and for both profiles; the lines of the graph paper will allow accurate placement of the protractor at each point.

When completed, you should be able to detect small ridges on the slopes of each profile, like those at B and D on Fig. 7.23 (1). These are known as beach **berms** and they result from increased wave activity at successive high tide levels. By comparing the two profiles, you should find that the berm at point H on profile 1 is probably linked to that on profile 2 (also point H). Since these profiles have been drawn to scale, a number of other measurements should be possible.

14 a) What is the height of the crest of each profile above point A?
b) Which is the higher of the two profiles, that to the north or that to the south?
c) At what height on each profile does the berm at point H occur?
d) Briefly comment on some of the difficulties and weaknesses of this kind of survey technique.

Other factors being equal, variations in the size of pebble are likely to influence beach height; the larger the pebble, the greater the height. Since our two profiles have revealed different heights, we need to establish whether any significant differences in pebble size exist between them. Of course, as Fig. 7.24 shows, there is likely to be a considerable variation in size at either location. We therefore need to take representative samples of pebbles from both sites and work out the

Coastal Processes and Associated Landforms

STORM BEACH PROFILE 1 AT 604928			STORM BEACH PROFILE 2 AT 603938		
Beach survey point	Angle of slope to next point in degrees + = up − = down	Length of slope section in metres	Beach survey point	Angle of slope to next point in degrees + = up − = down	Length of slope section in metres
seaward side			seaward side		
A	+14	2.92	A	+15	2.30
B	+ 9.5	1.53	B	+23	1.54
C	+16	1.45	C	+13	0.77
D	+10	0.64	D	+34	0.46
E	+22	1.90	E	+14	1.72
F	+18	0.85	F	+11	1.28
G	+24	1.45	G	+21	1.47
H	+ 3	0.85	H	+ 4	1.90
I	+18	1.26	I	− 2	1.15
J	+26	1.48	J	+ 4	1.28
K	+12	0.85	K	+14	0.43
L	−16	1.90	L	+ 6	1.00
M	−12	6.80	M	− 8	1.46
landward side			N	−15	1.23
			landward side		

Table **7.1** Storm beach profile data for Borth, West Wales

mean (average) size at each. The length of pebble long axis is a convenient measure of pebble size, as shown in Fig. 7.25.

Sampling of pebbles can prove difficult, but if stones are selected randomly from the same relative positions on each profile (e.g. the base, the crest or the same berm), inconsistencies can be minimised.

15 a) Write a brief account of the difficulties you might experience in selecting a *random* sample of pebbles from a storm beach.
b) Discuss ways *you* might choose to overcome such difficulties.

Fig. **7.24** Pebbles on the crest of the Borth storm beach

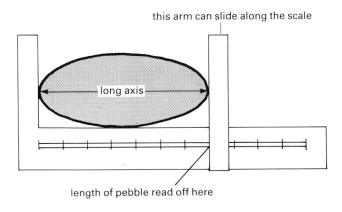

Fig. **7.25** Measuring pebble size

Tables 7.2 (a) and (b) present the results of pebble sampling on the Borth storm beach at the locations where the cross-profiles were drawn. Every effort was made to ensure that pebbles of the same rock type were measured. (Why?)

Coastal Processes and Associated Landforms

Sample 1	Long axis measurement (cm) of 40 randomly selected pebbles			
15.5	3.9	12.8	16.5	1.1
18.6	4.9	12.2	13.4	13.6
12.0	5.2	10.0	9.9	17.4
19.0	4.8	26.2	9.9	17.8
11.0	11.7	14.1	16.5	13.2
12.5	10.6	13.0	13.8	15.0
10.4	13.1	12.0	15.2	9.9
9.0	16.5	8.8	16.0	19.0

Table 7.2 (a) Pebble samples from profile 1 (604928) on the Borth storm beach

Sample 2	Long axis measurement (cm) of 40 randomly selected pebbles			
2.1	5.0	7.6	14.4	16.0
3.4	9.9	8.4	7.5	24.6
4.5	9.8	10.2	20.5	15.4
3.7	11.2	16.0	20.5	13.5
5.8	5.7	13.4	19.0	16.8
7.4	3.5	15.4	20.2	12.0
14.5	5.9	11.6	18.7	18.5
4.7	14.4	12.8	19.8	8.0

Table 7.2 (b) Pebble samples from profile 2 (603938) on the Borth storm beach

of the sample on the diagram (in Fig. 7.26 (a) the 20 largest pebbles) can be divided from the lower half by the **median** or 'middle value'. Each half can then be divided in two by **quartiles**, the upper quartile for the top half, the lower quartile for the bottom half. These are shown on Fig. 7.26 (a). The relationship between the medians and quartiles of the two samples will help determine the statistical significance of any difference. Fig. 7.26 (b) shows how this significance can be determined.

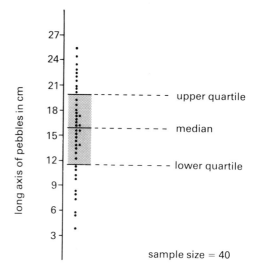

Fig. 7.26 (a) Typical dispersion diagram showing median and quartiles

16 a) Set up your own simple hypothesis, bearing in mind that we are trying to establish whether pebble size and beach height are related.
b) Using the data in Tables 7.2 (a) and (b), calculate the mean pebble size for each sample (add up the individual pebble sizes and divide the total by the number in the sample, i.e. 40).
c) Which sample has the greater mean pebble size?
d) Do you accept or reject your hypothesis?

17 a) On a piece of graph paper draw two vertical lines about 10 cm apart. Each line should be about 27 cm long. Mark a scale on each line, making sure the two scales coincide (e.g. 1 cm on the graph = 1 cm of pebble size). Label each line as in Fig. 7.26 (a), identifying them as samples 1 and 2 respectively.
b) Carefully plot the data from Tables 7.2 (a) and (b) on the appropriate lines to complete two dispersion diagrams. Draw in the medians and quartiles for each.
c) Using Fig. 7.26 (b) as a guide, how significant statistically is the difference between the two samples?

Your results will have revealed a difference between the mean pebble sizes at the two locations, but it is necessary to establish whether or not this difference is big enough to be regarded as **statistically significant**. For this purpose each sample can be regarded as a **statistical population**. Each can be plotted on graph paper in the form of a **dispersion diagram** (Fig. 7.26 (a)), the pebbles being accurately arranged in order of size. The top half

If the difference is shown to be significant, this provides clear evidence of **pebble sorting** or **grading** along the storm beach. A difference which is not significant may show i) that there is no sorting, or ii) that there was something wrong with the sampling method, e.g. sites too close together or pebbles not selected randomly enough.

Probably the simplest explanation for pebble sorting,

Coastal Processes and Associated Landforms   177

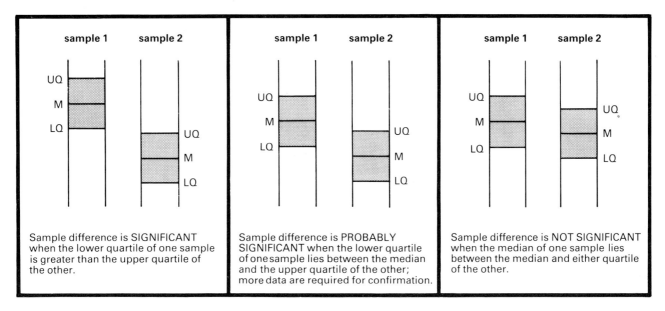

Fig. 7.26 (b) Method of using medians and quartiles to find how statistically significant a difference between samples might be

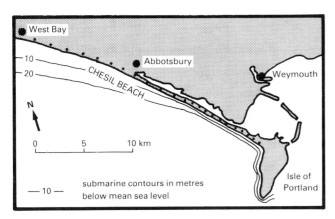

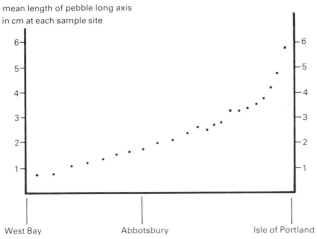

Fig. 7.27 Pebble sorting along Chesil Beach, Dorset

therefore expect the pebbles at Twyni Bâch to be smaller than those at Upper Borth on the Borth storm beach.

One of the best examples of pebble sorting can be seen at Chesil Beach in Dorset. This is a very long beach extending from West Bay to the Isle of Portland. It increases progressively both in height and width from west to east; at the Isle of Portland the beach is over 13 m above mean sea level compared to only 7 m at Abbotsbury. At each of 23 sites along the beach, geomorphologists took samples of 500 pebbles. The results are shown in Fig. 7.27.

18 Study Fig. 7.27 carefully.
a) Measure the total length of the beach.
b) Describe, quoting pebble sizes where appropriate, how effective pebble sorting appears to have been.
c) By referring to an atlas, establish the direction of longshore drift on Chesil Beach.
d) In view of our remarks above, is there anything unusual about the relationship between pebble size and the direction of longshore drift? Justify your answer.

when it occurs, is that as pebbles are being transported by longshore drift, they are going to be steadily worn down in size. *Thus the further from their source of origin pebbles drift, the smaller they become.* We would

You will have realised by now that coastal processes cannot always be easily explained. Pebble sorting at Chesil Beach is not fully understood. One suggestion is that the steeper offshore gradient towards the Isle of Portland (look at the submarine contours on Fig. 7.27) allows waves to approach closer to shore before breaking up. By retaining more energy, waves can throw bigger pebbles higher up the beach.

Depositional Coastlines

Spits and Bars

Storm beach development and longshore drift can produce some very distinctive coastal landforms. The most significant of these is the **spit**, a shingle ridge which projects out from the coastline. As the storm beach gradually extends from the land, it acts as a backbone against which other features such as sand dunes and salt marshes can grow. The storm beach at Borth is a classic spit. We can use our findings so far to show how it developed.

> **19** Study carefully map extract 2.
> a) In your exercise book, draw a large outline of the coastline shown on the extract.
> b) Identify that part of the coastline where erosion is taking place. Look for wave-cut platforms and cliffs and mark them on your map. (See Fig. 7.16 for symbols you can use.) Try to quote cliff heights where possible (e.g. squares 5986 and 5987). Annotate your map accordingly.
> c) By means of an arrow, mark on your map the direction of longshore drift, which you established in exercise 12.
> d) The spit extends into the Dovey estuary. Locate this in your atlas. Suggest a reason why the spit is unlikely to seal off the estuary completely.

When the direction of the dominant waves is significantly different from the direction of longshore drift, then the spit may be curved at its end. Such a spit is known as a **recurved** or **hooked spit**. Fig. 7.28 shows how this might happen – it is a **model of recurved spit development**.

An actively eroding cliff supplies the material from which the spit grows. The actual shape of the spit is controlled by the extent to which powerful dominant waves interfere, sweeping material away from the line of longshore drift to form **laterals**. A spit will lengthen considerably if there are long periods when dominant waves are relatively weak. The time scale involved tends to be in hundreds, rather than tens of years. Note that as the cliff becomes sheltered by the spit, it starts to degrade; that is, its profile becomes less steep because rainwash and slumping on the cliff face have more influence than wave undercutting (see Fig. 7.9 IV).

A good example of a recurved spit is that at Blakeney in Norfolk, shown in Fig. 7.29. The spit is composed of material eroded from very weak cliffs of mainly glacial deposits to the east.

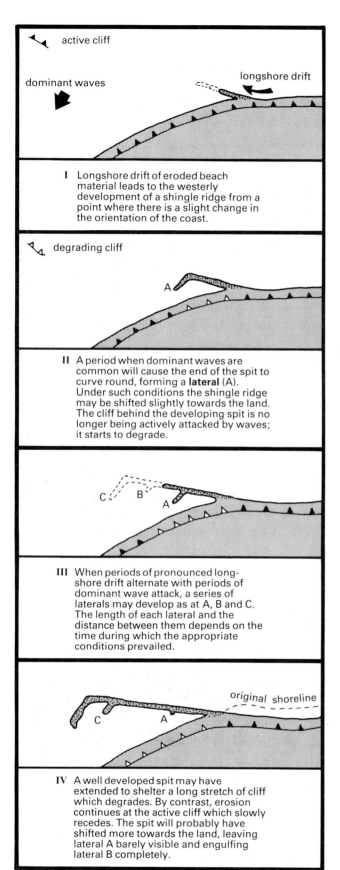

Fig. **7.28** A model of recurved spit development

Coastal Processes and Associated Landforms

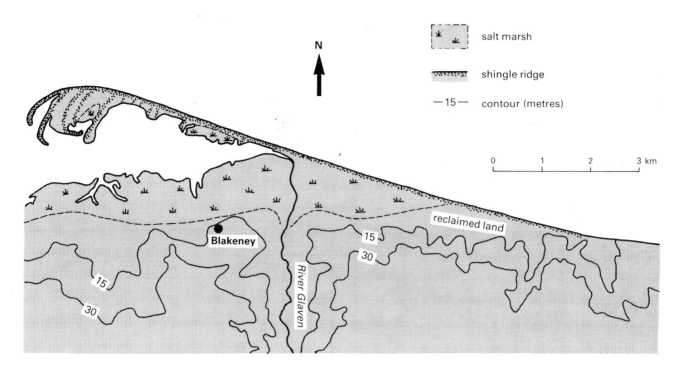

Fig. 7.29 Blakeney Spit, Norfolk

20 a) Study Fig. 7.29 and draw the outline of the coast in your exercise book.
b) Locate Blakeney in your atlas and establish the dominant wave direction. Mark it on your outline.
c) What is the approximate length of the spit?
d) Identify on the map and mark on your outline where you would expect to find i) a degrading cliff-line, and ii) actively eroding cliffs. State their heights and compare their gradients.
e) How would you describe the conditions that have *recently* controlled the shape of the spit? Justify your answer.
f) Mark on the spit a stretch developed under a period of active elongation (longshore drift).
g) Briefly compare and contrast the conditions influencing the growth and development of the Borth and Blakeney spits.

When a shingle ridge extends from the coast to link up with an island just offshore, a **tombolo** is formed. Chesil Beach, considered above, is a fine example. However, it is a far more complicated feature than might be realised, particularly since (as you should have noticed in exercise 18) the pebble sorting is in the opposite direction from what we would expect!

One theory for the formation of Chesil Beach is that it developed as an offshore shingle ridge which then moved towards the shore. Such ridges are known as **offshore bars**. Initially they have no link with the coast and are probably formed by waves breaking some distance offshore. A ridge of material steadily builds up, rising eventually above high tide level. A gentle offshore gradient is therefore an important condition for the development of offshore bars, since water depth

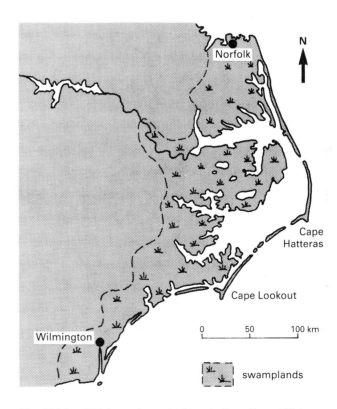

Fig. 7.30 Offshore bars in North Carolina, USA

controls the point at which waves break. Some of the finest examples, many kilometres in length, are found off the coast of North Carolina in the eastern United States; they are shown in Fig. 7.30. The sheltered water behind an offshore bar is referred to as a **lagoon**.

On rare occasions, a whole series of shingle ridges may develop next to one another, leading eventually to the formation of a triangular shaped projection of land; this is known as a **cuspate foreland**. The classic British example is that of Dungeness in Kent, shown in Fig. 7.31. You should note the alignment of the shingle ridges and the huge area of Romney Marsh, now reclaimed and used as farmland. The whole feature has probably taken 3000–4000 years to reach its present shape, with marked changes in the last few centuries.

---

21 By reference to Figs. 7.30 and 7.31 answer the following.

a) *For Fig. 7.30*: Measure the length of some of the offshore bars and their distance from the present coastline. What evidence might you look for in the region to show that this present coastline has not always existed?

b) *For Fig. 7.31*: Calculate the approximate area of the cuspate foreland in front of the original cliff-line. Estimate the percentage of this area made up from shingle ridges. Use the method shown in Chapter 5 for measuring the size of an irregular area (see p. 107).

c) Work out the mean annual rate of advance of the Dungeness coastline from the information given on map B. Compare your figure with the mean annual rate of retreat calculated for the Holderness coast in exercise 10.

d) By reference to the submarine contour offshore from Dungeness, suggest why the future rate of advance of the cuspate foreland into the English Channel will be much slower than in the past.

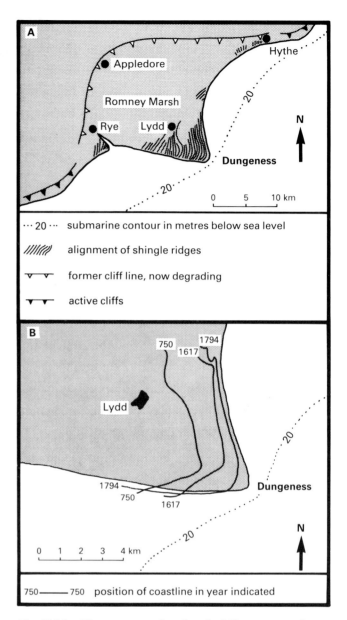

Fig. **7.31** The cuspate foreland of Dungeness in Kent

Salt Marshes and Sand Dunes

Shingle spits provide shelter from wave attack and therefore the conditions are ideal for the development of a **salt marsh**. Note the presence of marsh land, including reclaimed marsh, in Figs. 7.29, 7.30 and 7.31. The calmer water encourages the accumulation of fine deposits which, as they build up, will eventually be colonised by **halophytic** plants. These can tolerate the salty sea water and they help to bind the salt marsh together. Salt marshes are usually characterised by a number of drainage creeks (see Fig. 7.32). In time, the sediments will reach a level where they will be submerged by fewer and fewer high tides. Because of this, the marsh water will become increasingly brackish, as fresh water begins to mix with salt water. Further sedimentation may cut off the sea altogether, giving a **fresh water marsh** which can be reclaimed.

A key feature of the Borth spit is the line of **sand dunes** which extends northwards from northing 91. When a gently sloping offshore profile exists, much land will be exposed between high and low tides. Because of this, some of the surface layers of the exposed sand may dry out sufficiently to be picked up by an onshore wind. These sand particles will be deposited just behind the shoreline, gradually building up to form sand dunes. The presence of vegetation, particularly **marram grass** with its long deep roots and its ability to survive on shifting sand, encourages dune growth. The tufts of marram grass in Fig. 7.33 are not large, but their importance to the stability of the dune is considerable.

Coastal Processes and Associated Landforms

22 Return to the map of the Borth coastline which you drew for exercise 19 and refer to map extract 2.
a) Identify on the extract areas of i) salt marsh, ii) fresh water marsh, and iii) reclaimed land used for grazing (what map symbol suggests this?). Mark these on your map.
b) Locate the degrading cliff-line and calculate the average gradient at two representative points.
c) Mark the two isolated pieces of slightly higher ground at Ynys-fergi (614897) and Gwynfryn Hall (644918). By reference to Fig. 7.12, can you suggest what they might be? Complete the map with relevant annotation for these points.

Fig. **7.32** Salt marshes near Blakeney, North Norfolk

Sand dunes may combine to form a **dune belt** which will stretch along and across the spit. Within the belt there will be a contrast between dunes on the seaward side and those further inland. Seaward dunes are much larger and more active than those inland which may have decayed altogether into **remanié dunes**. Differences in the amount of sand supply from the foreshore help to explain these variations in size, as well as influencing the degree and type of grass coverage. Removal of the dunes' vegetation cover may allow wind erosion to create a bare dune hollow or gulley, known as a **blow-out**.

Fig. **7.33** Marram grass

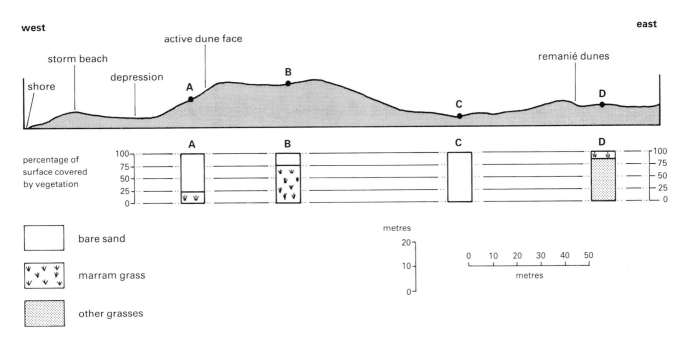

Fig. **7.34** West–east profile of Borth Spit along northing 937

Coastal Processes and Associated Landforms

The Effect of Sea Level Changes on Coastal Landforms

Sea levels are rarely stable for long periods of time, largely because the world's climate is constantly changing. As it cools in a glacial period for example, more of the world's water will be stored in the form of ice sheets, ice caps and glaciers, leading to a fall in sea level (base level); the opposite will occur during an interglacial period. Land levels may also change. Earth movements can push land up from beneath the sea, whilst the melting of an ice sheet removes a great weight from the Earth's crust and allows it to rise. Parts of Scandinavia, for example, are still rising at an average of one centimetre a year. Such changes in sea and land levels may lead to the development of some very distinctive coastal landforms, as well as helping to explain some of those we have already discussed. For example, the development of an offshore bar above the high tide level can be explained more easily if we state that it is accompanied by a slight fall in sea level.

Two major types of relative movement can be identified. A *positive movement* of base level occurs when the sea rises in relation to the land, causing **submergence**. By contrast, a *negative movement* will lead to **emergence**. Coastlines are frequently very complicated, often showing signs of *both* types of movement, though one generally stands out more clearly.

Coastlines of Submergence

A distinction can be made between lowland and highland coastal submergence. In lowland areas the rise

Fig. **7.35**   Sand dunes on Borth Spit

Fig. 7.34 shows a profile across the Borth spit dune belt, drawn using data from a clinometer survey. Notice the depression immediately behind the storm beach; this is visible in Fig. 7.21. Fig. 7.35 shows two photographs of dunes taken along the profile.

> 23   Study Figs. 7.34 and 7.35 carefully.
> a) Write a description of the differences between the active and remanié dunes. You should i) measure dune heights, ii) measure angles of slope, and iii) contrast the amount and type of vegetation at points A, B and D.
> b) How would you explain the feature at C?
> c) Which of A, B, C or D are represented by I and II in Fig. 7.35? Justify your answer.

Fig. **7.36**   The mouth of the River Solva, South Wales: an example of a ria

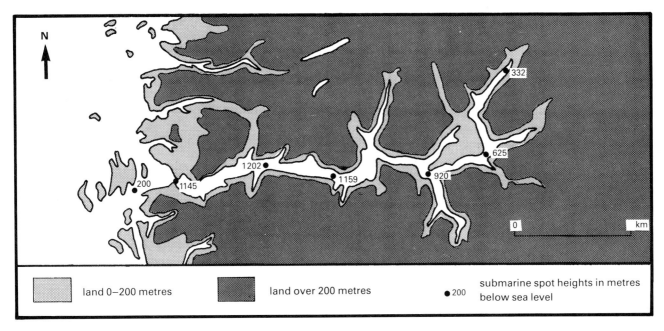

Fig. **7.37** A coastline of submergence

in sea level produces very broad and extensive inlets known as **estuaries**. The gentle offshore gradients which result are ideal for the development of offshore bars and salt marshes. In highland regions the drowning of the lower course of a river valley will produce a **ria**. This is an inlet characterised by many branches which are former river tributaries. The former river channel may still be detected in the submarine contours. The slopes of the ria will run down right to the water's edge after submergence, though eventually wave attack will cut a small cliff into them, as shown in Fig. 7.9; these features are clearly visible in Fig. 7.36. When glaciated U shaped valleys are submerged, **fjords** are produced. They are much more steep-sided than rias and their branches tend to be more at right angles, reflecting the position of former hanging valleys. Fjords may be extremely deep, the result of glacial overdeepening, though their mouths are very much shallower.

Figs. 7.37 and 7.38 are two maps of submerged coastlines, one of Milford Haven, a ria in South West Wales, the other of the Sogne Fjord in Norway. You will note that no scale has been given on either map.

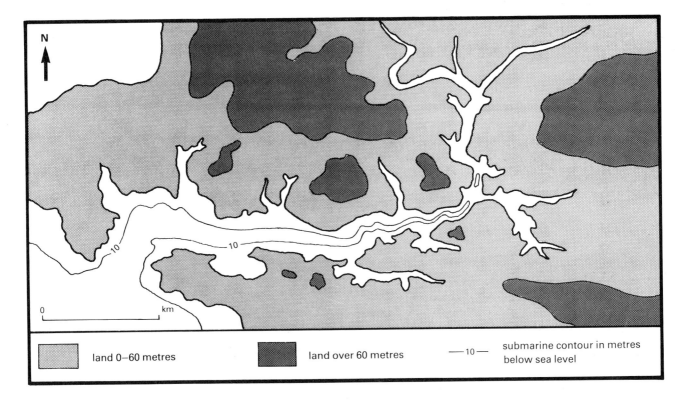

Fig. **7.38** A coastline of submergence

24 Study Figs. 7.37 and 7.38 and re-read the paragraph above carefully.
a) Write a description of each coastline and identify which location is represented by which map, using the information on the maps. State the reasons for your choice.
b) When this is complete (not before!) locate the two features in your atlas and select a suitable scale for each map.

One of the interesting features of coastlines of submergence (particularly lowland coasts) is the **submerged forest**. If a rise in sea level swamps a forest area, the trees will die but the roots and stumps may be preserved in a peat layer, protected by accumulating silt and sand deposits. Later, destructive wave action may erode these deposits to expose the tree stumps below. These remains can be dated by modern techniques.

A good example of a submerged forest can be found between northings 91 and 92 on the foreshore of the Borth spit. It is shown in Fig. 7.39 and dates from about 8000 years ago when it was probably part of a river's flood plain or a coastal plain. The layer of peat is clearly visible in photograph B, surrounding the old tree stump.

25 a) Draw a sequence of diagrams to explain how a submerged forest might be formed.
b) Return to the map of the coastline near Borth which you used for exercises 19 and 22. Mark on the submerged forest.
c) Using all the information you now have on the Borth coastal region, write an account of the likely sequence of changes occurring along the coastline from about 10 000 years ago to the present day. Include sketch maps to indicate what the coastline might have looked like at different times.

Fig. **7.39** The submerged forest near Borth

Coastlines of Emergence

We mentioned above how the formation of offshore bars might be assisted by a slight relative fall in sea level, causing emergence. In fact the entire coast of the eastern USA is one of emergence.

On a more local scale, perhaps the best indication of a relative fall in sea level is the presence of a **raised beach**. Fig. 7.40 shows how this might develop following on from Fig. 7.9 III.

In time, the new cliff may recede so much that all evidence of the raised beach is eroded. Alternatively, there may be further relative falls in sea level, so that the raised beach becomes higher and higher. In this case the term **marine terrace** or **marine platform** is used, but detection in the landscape may be difficult because of the effects of the 'normal' processes of denudation. However, a marine terrace does exist at 195 m above sea level in South East England, probably dating from the early years of the Pleistocene period (see Chapter 6).

Fig. 7.41 shows a fine example of a raised beach.

26 a) Study the photograph carefully and identify the key features shown in Fig. 7.40.
b) Draw a fully annotated sketch of the photograph, explaining the processes that led to the formation of the features shown.
c) Can you suggest why there should be good examples of raised beaches off the west coast of Scotland? Fig. 6.5 (b) will give you a clue.

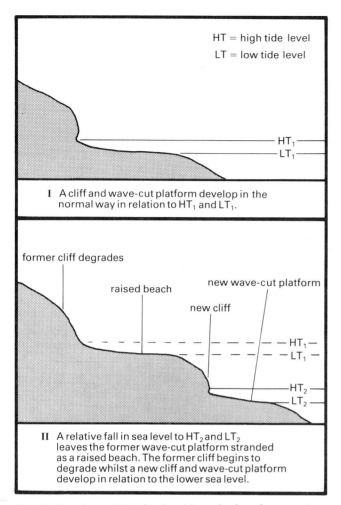

Fig. **7.40** A model of raised beach development

Classification of Coasts

There are a number of ways of classifying coasts. The method we will look at relates coastal orientation to the structure of the landscape. Under this system two main classes of coast exist. When the coastline runs parallel to the main structure of the landscape, as in Fig. 7.17, a longitudinal or **concordant** coast results. When the coastline cuts across the structure there will be an alternation of headlands and bays, producing a transverse or **discordant** coast, as in Fig. 7.16. Concordant and discordant coasts may also be known as Pacific and Atlantic types respectively. You will see from Fig. 7.42 that many of the world's coasts can be classified in this way; you will also appreciate how the terms Pacific and Atlantic arose.

27 **a)** Using your atlas, draw sketch maps of the coasts of i) Yugoslavia, and ii) South West Ireland. Say which type of coast each represents, justifying your answers. The trend of the relief in both these regions gives a clue to the structure of the landscape.
b) Review quickly the contents of this chapter and suggest other ways in which coasts could be classified.

Fig. **7.41** A raised beach

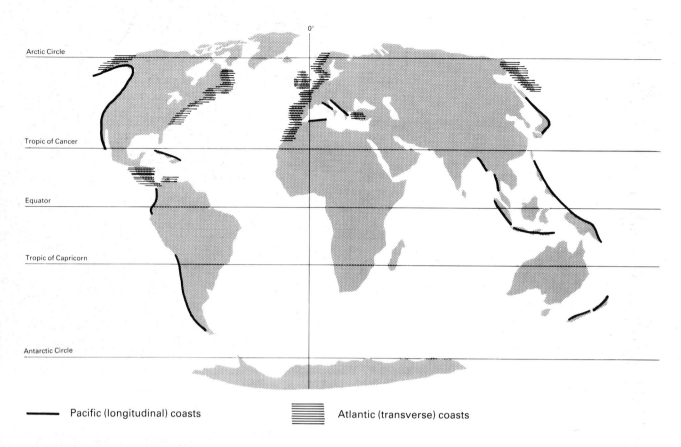

Fig. **7.42** World concordant and discordant coastlines

Coastal Processes and Man

Coastal erosion and slumping (see Figs. 7.18 and 2.7 respectively) pose a constant threat to man. Coastal deposition also creates difficulties. Siltation is frequently a problem in our ports today, particularly with the ever increasing size of vessels.

People have tried to tackle these problems in a variety of ways. To combat sea erosion, sea defences have been built in places, but these may be prohibitively expensive. In some cases, therefore, boulders have simply been dumped on the foreshore to provide extra protection fairly cheaply, as well as making up for losses through erosion. Groynes slow down the rate of longshore drift of material, but often at the expense of increasing erosion further along the coast which, deprived of beach material, will become more exposed to wave attack. Finally, siltation in ports is only relieved by constant and expensive dredging.

However, coastlines may offer many benefits to mankind. Rias provide deep natural harbours which are of great economic importance to trading countries; Milford Haven for example, is a leading oil port. The recreational importance of coasts has increased in recent years. To protect some attractive coastlines from over-development they may be designated **National Parks**, as in the case of the Pembrokeshire coast, or **Areas of Outstanding Natural Beauty**, as in the case of much of the coastlines of Cornwall and Dorset.

But we must not forget that today, man himself is a major geomorphological agent. For example, the sand dunes at Borth are under great threat from the effects of too many trampling holidaymakers' feet. Severe erosion and blow-outs have resulted and this has meant urgent remedial action, some of which can be seen in Fig. 7.43.

Fig. **7.43** Sand dune conservation near Borth

Coastal Processes and Associated Landforms

> **28 a)** Study Fig. 7.43 and make a list of the measures that appear to have been taken to try to prevent the erosion of the dunes.
> **b)** Discuss and list other ways in which coastlines have been affected by man.

Workshop

1 Study Fig. 7.44 which was taken at 586843 on map extract 2. Describe in detail the features that are shown and the processes responsible.

Fig. **7.44** Shoreline to the north of Clarach

2 **a)** The southern tip of the Holderness coast examined earlier in this chapter (see Fig. 7.18) is Spurn Head. This is shown in Fig. 7.45. Locate Spurn Head in your atlas, say what the feature is and describe in detail the coastal processes which have led to its development.
b) Do the same for Scolt Head on the North Norfolk coast, shown in Fig. 7.46.

3 The term 'model' has been used on a number of occasions in this chapter. Refer back to Figs. 7.9, 7.12, 7.28 and 7.40, and in your own words try to explain the main purpose and usefulness of a model in the understanding of landform development. Use the work done in the last part of exercise 9 to build up a model of concordant coastline development.

4 Look through some of your old 'seaside' holiday photographs or slides. See how many of the coastal features discussed in this chapter you can recognise. Describe the processes which may have been important in their formation.

5 Refer back to the slope sampling exercise you undertook in Chapter 2, comparing the steepness of slopes on the two map extracts (exercise 11). Following the method outlined in exercise 17 in this chapter in conjunction with Fig. 7.26, use the results you obtained earlier to establish whether or not there is any significant difference in steepness between the two slope samples.

Fig. **7.45** Spurn Head, Lincolnshire

Fig. **7.46** Scolt Head, North Norfolk. *Crown Copyright Reserved*

Desert Processes and Associated Landforms 8

We considered in Chapter 6 how the 'normal' processes of landscape evolution, i.e. those involving running water, might be considerably modified or replaced during a period of glaciation. Normal processes are also going to be modified in areas where there is little or no running water, resulting in a rather different set of landforms. Such landscapes can be broadly described as **deserts**.

The terms 'desert' and 'semi-desert' conjure up a variety of different images in our minds. Endless sand dunes, vast stony plains and the typical 'cowboy and indian' country with its cacti and canyons are probably some of the first landscapes we tend to think of. What they show is that there are many different types of desert landscapes and this variety suggests the involvement of several processes. Deserts are not straightforward environments.

A simple definition of deserts is not easy. Some, like the Atacama, are very hot and on the coast, whilst others, like the Gobi, can be extremely cold and deep in a continental interior. Nevertheless, there are certainly some common features. There is a general absence of surface water, which is largely responsible for the lack of vegetation. Chapter 4, which touched on the causes of deserts, showed that the excess of evaporation over precipitation, producing **aridity**, is a key factor in desert formation.

About one third of the world's land surface is desert, with 4% regarded as extremely arid, 15% as arid and 14.6% as semi-arid. The location of these three zones of aridity is shown in Fig. 8.1.

> 1 Study Fig. 8.1 in relation to your atlas.
> a) On an outline map of the world, mark on the 'arid' and 'extremely arid' zones.
> b) Use the work done in exercise 41 of Chapter 4 to name the world's hot deserts. Use your atlas to identify some of the world's cold deserts. (You may wish to return to p. 87 for a definition of hot and cold deserts.)
> c) Turn to Table 4.3 and the section in your atlas on climatic stations, if it contains one. Identify one or two suitable stations which are located in the aridity zones you have marked on your map.
> d) Using the data given for each station, describe their climates and compare them with the description of desert types given in Chapter 4. Pay particular attention to mean annual rainfall totals and any seasonal distribution, as well as differences in mean monthly temperatures.

You should have noted that even in the most 'extremely arid' regions, rainfall still occurs, though *actual* totals are likely to vary considerably from one year to the next (see Fig. 4.41). However small and intermittent rainfall might be, one must expect to find evidence of drainage systems, even though they may be occupied for only brief periods. Thus deserts usually contain evidence of the work of running water.

In Chapter 6 we mentioned briefly how wind action alone is able to redistribute glacial deposits in the form of loess, whilst in Chapter 7 we saw how wind has been responsible for the build-up and destruction of coastal dune systems. In both circumstances the effectiveness of the wind would have been much greater, the drier the deposits. A simple experiment allows us to confirm this. If you direct air from a hairdryer onto two trays of identical coarse-grained sand, one wet, the other dry, the dry sand will shift more easily. This is because air can penetrate between the dry sand grains, whilst the wet grains are held together by the water droplets. Thus, with little or no moisture, deserts should contain some evidence of the work of wind.

Fig. 8.2 shows a large area in southern Algeria which is part of the central Sahara.

> 2 Study Fig. 8.2 and try to locate the region in your atlas by using the lines of latitude and longitude shown.
> a) From Fig. 8.1, what kind of aridity conditions would you say exist in the region?
> b) What evidence on Fig. 8.2 is there of the work of running water and wind?

In fact, in the region shown, some of the more distinctive desert landscapes are represented, though by no means all. The rugged upland region of the Ahaggar massif is further broken up by the deeply-incised temporary water courses known as **wadis**. Surrounding the Ahaggar are broad areas of rocky and stony desert, known in the Sahara as **hammada** and **reg** respectively. Reg is the most common desert landscape in the Sahara; the Plateau du Tademait and the Tanezrouft plain are good examples. Low-lying areas often contain salt flats or **playas**, which result from the rapid evaporation of temporary lakes; Sebkra Mekerrhane is typical of this type of landscape. Finally there are the vast areas of sand accumulations known as **ergs**. In fact one fifth of the Sahara is covered with wind-blown sand in 25 large ergs. If spread evenly over the entire 9 000 000 km^2 of the Sahara, the thickness of the sand within the ergs would average 15 metres!

Desert Processes and Associated Landforms

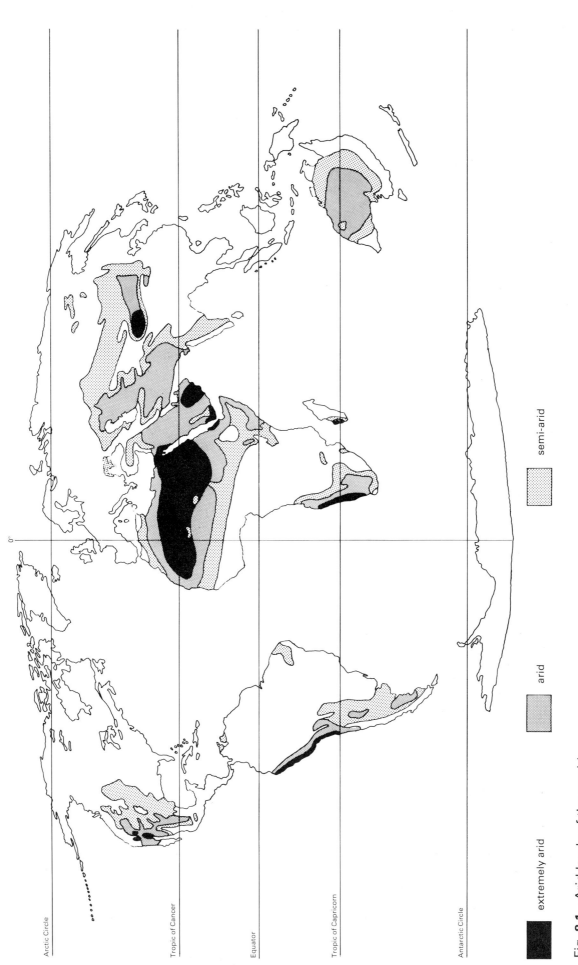

Fig. **8.1** Arid lands of the world

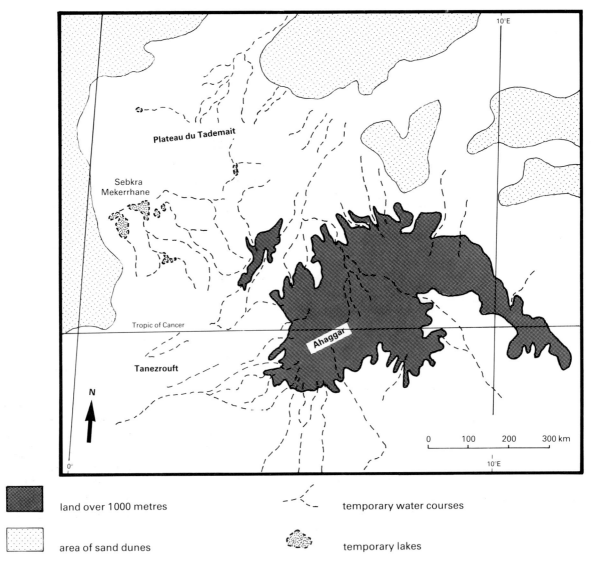

Fig. **8.2** The Sahara Desert in Southern Algeria

3 a) On an outline copy of Fig. 8.2, using suitable symbols, mark the areas representing the types of desert landscape mentioned above.
b) From data given above, calculate the volume of wind-blown sand in the Sahara.
c) Examine in your atlas some of the other deserts of the world (e.g. the Kalahari in southern Africa and the deserts of central Australia) and look for evidence of the landscapes so far described.

All these landscapes, to a greater or lesser extent, owe their distinctive features to the geomorphological agents of running water and wind. The activity of these agents is accentuated by the lack of vegetation. But for either agent to operate effectively, there must first be an initial bedrock load. We must therefore examine weathering processes in deserts.

Desert Weathering

In Chapter 2 we saw that there are two main types of weathering process, mechanical and chemical. Because of the lack of water in deserts it was once assumed that chemical weathering was comparatively rare. Rather, it was thought that the great diurnal range in temperature was directly responsible for the mechanical breakdown of the rocks. Cloudless desert skies mean rapid heating of the land in the day followed by rapid cooling at night, as the heat is radiated back into the atmosphere. It was believed that this alternate heating and cooling caused stresses in the surface of the rocks as they expanded and contracted, eventually leading to their disintegration – a process known as **exfoliation**. Some laboratory experiments were undertaken to try to show this:
Experiment A A piece of granite with a polished surface was heated rapidly by an electric heater and cooled by dry air. The temperature change involved was 110°C. This was repeated 89 100 times, at the end of which there had been no visible breakdown of the rock, even under microscopic examination.

Experiment B This was identical to experiment A, except that this time the cooling was done by a fine spray of water. After only 900 repetitions the rock surface showed clear signs of cracking; the rock had certainly lost its polish.

> **4 a)** How much greater was the temperature range of the experiments compared to the typical diurnal range in a desert?
> **b)** For the equivalent of how many years did each of the experiments simulate the diurnal range of temperature?
> **c)** From the results, i) what conclusions can we draw about the likely importance of exfoliation *as described above* in desert weathering, and ii) are our assumptions about the importance of chemical weathering valid? Justify your answers.

It is worth stating that experiments such as these do not *prove* that mechanical exfoliation is unlikely, even under such extreme conditions. Natural conditions cannot easily be reproduced in the laboratory. One major difference was that in the above experiments, the rock was free to expand and contract, whereas in reality, as part of a much larger rock massif, this would not be the case. Another point worth considering is that our assumed definition of exfoliation might be inadequate; perhaps the presence of water *is* necessary for the process after all.

However, if chemical weathering does occur, we must ask where the necessary moisture comes from. During the day the relative humidity (see Chapter 3) in a desert may be as low as 25%. The sharp night-time fall in temperature increases relative humidity so that dew eventually forms. Although it is quickly evaporated the following morning, it is believed that such dew is important in desert weathering. In the colder deserts like the Gobi, this moisture will allow freeze-thaw weathering to be quite common.

Weathering in deserts will attack points of weakness in the rocks, like joints and bedding planes. This produces a surface which may be covered by coarse and angular shaped boulders and stones. The Sahara contains extensive exposures of sandstone rocks and resistant volcanic rocks like those of the Ahaggar. Since desert sand is mainly composed of quartz and since quartz is one of the main minerals in sandstone and granite (see Chapter 1), it is easy to appreciate how the vast quantities of sand you calculated in exercise 3 might have originated. The load produced by weathering will be used in the process of erosion.

We must now consider the influence that running water and wind have as agents of erosion in the formation of desert landscapes. Whilst they may operate in conjunction with one another, with some landforms bearing the imprint of both, for convenience we shall consider them separately.

Running Water in the Desert

Desert rainfall is likely to show a strong convectional character, resulting in torrential downpours. Sometimes as much as 300 mm may fall in a single day, often followed by several years without any rain. The raging torrents of water resulting from such rainfall are known as **flash floods**, sometimes sub-divided into **stream floods**, which occur in pre-existing water courses, and **sheet floods**, extensive flows of water over the entire desert surface.

Flash floods are capable of much geomorphological work, especially when confined as stream floods. The great amounts of debris produced by weathering during the years without rain are easily picked up by the water torrents. The absence of vegetation certainly makes this easier.

You will recall from Chapter 5 that when rain falls it may drain across the land's surface as run-off (overland flow) or percolate through the ground, emerging eventually as base-flow or through-flow (see Fig. 5.12). The main factors influencing run-off are the permeability of the bedrock and the height of the water table, the absorption capacity of the soil (itself affected by how much moisture it is holding) and the levels of evaporation and transpiration.

> **5 a)** Using the factors mentioned above as a guide, return to Fig. 8.2 and suggest reasons why the wadi systems seem i) not to be completely integrated (like the streams shown, for example, in Fig. 5.5), and ii) to totally disappear well before reaching the sea.
> **b)** Redraw Fig. 5.12, adapting it to the desert conditions which have so far been described.

We can build up a picture of a typical desert drainage system. It should be realised that only about 10% of the already small amounts of desert rainfall actually results in surface run-off (why?). Since rainfall is more likely to occur in upland regions (why?), it is here that our examination should begin. Indeed a close study of the Sahara region in Fig. 8.1 will show that the Ahaggar massif featured in Fig. 8.2 is a 'semi-arid' region compared to the 'extremely arid' zone immediately to the north. The torrential nature of the rainfall means that run-off will be confined to a wadi as a stream flood. When it emerges from the uplands it may flow across a

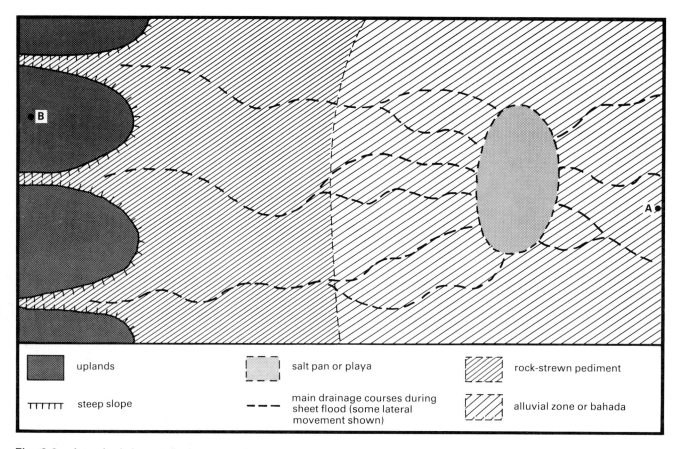

Fig. 8.3 A typical desert drainage system

rocky plain as a sheet flood, before disappearing by percolation or evaporation, or ending up as a temporary lake. Such a system is shown in Fig. 8.3, with the cross-profile from B to A being represented by Fig. 8.4.

By combining these two diagrams, some of the main components of a typical desert landscape can be identified. In time the upland massif may be worn back sufficiently by stream action to leave an isolated 'island mountain' or **inselberg**. The inselberg rises steeply from the surrounding gently sloping rocky plain or **pediment**. The pediment eventually merges into a zone of deposition called the **bahada** which may contain salt pans or playas. Collectively, the pediment, bahada and playa are known as the **piedmont**.

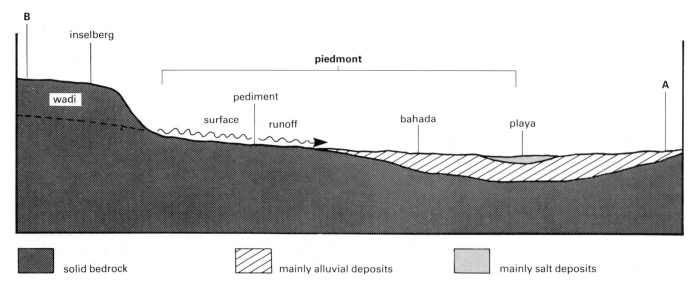

Fig. 8.4 Cross-profile of a desert drainage system

Desert Processes and Associated Landforms

6 a) Both the wadi and rock pediment are features produced by erosion. Refer back to Chapter 5 which shows how erosion by running water takes place. Why, when it occurs, is water erosion so rapid in deserts?
b) Why should rocks and boulders litter the pediment?
c) The bahada and playa are zones of deposition. Where do the deposits forming the bahada originate?
d) What atmospheric processes encourage the development of playas?

Fig. **8.5** A wadi in Morocco

Fig. 8.5 shows a typical wadi in an upland region in Morocco. The parched, vegetation-free nature of the valley is clearly seen, but in times of flash flooding such wadis may become very dangerous places and drownings are not uncommon.

7 a) Study Fig. 8.5 carefully and in your exercise book write a detailed description of the wadi. You should examine in particular i) the shape of the wadi sides and bottom, and ii) the scale of the features, using the people in the photograph as a rough guide. Where possible you should draw representative diagrams to illustrate the features.
b) Refer back to Chapter 5 (p. 117) and contrast the shape of a 'normal' river valley with that of the wadi. What processes would appear to be much more active in shaping the valley sides in a 'normal' environment than in a desert?
c) How would you explain i) the rock rubble on the wadi floor, and ii) the numerous gulleys on the wadi side?

the **mesa**, **butte** and **pinnacle**. In the early stages a **canyon** may exist. Canyons are more pronounced than wadis, with permanently flowing, rather than temporary, streams which have their sources outside the arid zones; such streams are known as **exotic streams** (or rivers). The Grand Canyon (shown in Fig. 8.24) is the classic example of this type of landscape; it has been formed by vertical erosion by the Colorado River.

Inselbergs occur in a variety of different forms. Perhaps the most striking are the round-topped inselbergs, like the famous Ayers Rock in Australia (shown in Fig. 8.25) and the flat-topped inselbergs, like those in Monument Valley in the United States (Fig. 8.6). A characteristic feature of all inselbergs is the abrupt change of slope angle where their fronts meet the surrounding pediment. Fig. 8.7 shows how they might have developed through the process of **backwasting**. This could be considered as a model of inselberg development. The resistant nature of the cap rock helps to protect the rock beneath it, but in time the inselberg will steadily erode through successive forms,

8 Study Fig. 8.6 carefully.
a) Identify examples of a mesa, butte, pinnacle and pediment.
b) Is there any evidence that running water has played a part in the development of these features?
c) How would you explain the large piles of rubble (scree) at the foot of the features shown? You may find it helpful to refer back to Fig. 2.1 (p. 24).

Fig. **8.6** Monument Valley, Utah, USA

The formation of the pediment, as Fig. 8.7 suggests, is closely allied with the development of the inselberg; as the inselberg front retreats, so the pediment extends. The pediment is an erosional feature in solid rock. Its development may be linked to lateral erosion by the floodwaters emerging from the uplands (see Fig. 8.3), but a key point at issue is whether sheet floods actually *result from* or *cause* pediments. Nevertheless, the pediment represents an important zone across which much eroded and weathered material is carried by running water, to be deposited in the alluvial bahada zone. The great energy of the sheet floods means that a very high concentration of sediment can be transported. As the floodwaters converge on the interior desert basin, high rates of evaporation cause the temporary lake, which has no outlet, to dry up. This leaves a salt flat or playa.

9 Turn to a map of Australia in your atlas.
a) In your exercise book draw an outline map of Australia and on it mark as many 'temporary' lakes as you can, naming the largest. These are all examples of salt flats. Measure the length of some.
b) From your atlas describe the relief of the area in which these salt flats are found. Annotate your map accordingly.

By now you may have suspected that the size and extent of some of the landforms considered are rather too great to have evolved solely under the influence of present day, infrequent run-off conditions. We saw in

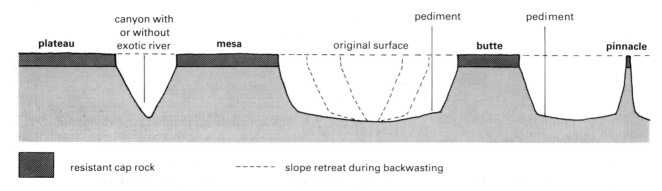

Fig. **8.7** Inselberg development

Chapter 6 how patterns of climate can change in a comparatively short time and another look at Fig. 6.3 (p. 136) will show that not all of today's deserts were deserts in the past. In fact many of the temporary water courses shown in Fig. 8.2, radiating out from the Ahaggar, were almost certainly formed at a time when rainfall was greater than today; their size is far too big for the present run-off they have to cope with. Such climatic wet spells are known as **pluvial periods** and the drainage systems shown probably resulted from several of these periods, the most recent of which ended more than 15 000 years ago.

> **10 a)** Refer back to Fig. 6.3 and the sections on climate and natural vegetation in Chapter 4. Using these, describe the likely conditions of climate and vegetation prevailing in the Sahara region 18 000 years ago.
> **b)** What were the causes of the changes that have since occurred?

Wind in the Desert

Wind, like running water, can perform a full range of geomorphological work in deserts, from erosion through to deposition. Loose, dry material on the desert floor can be picked up by desert winds, forming either **dust storms**, which may extend to heights of several kilometres and blow over hundreds of kilometres, or **sand storms** which rise no more than two metres from ground level. The removal of loose material in this way is known as **deflation**. Perhaps more important is the **abrasion** that follows when wind-borne particles hit exposed rock, causing it to wear down.

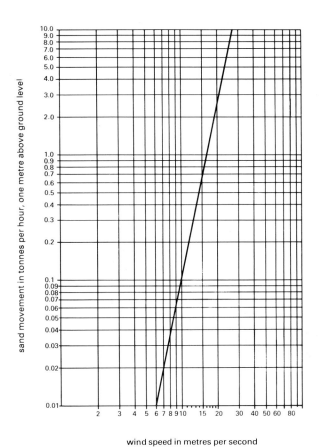

Fig. **8.8** The effect of wind speed on sand movement

Fig. 8.8 is a graph showing how effective wind can be at moving sand. Notice the scales on the graph are logarithmic and not arithmetic. On the graph, sand movement is being measured at a height of one metre above ground level. For comparison, 1 m/sec equals 3.6 km/hr.

1 Wind speed (metres per second)	2 Amount of material moved (tonnes per hour)	3 Wind speed doubled	4 Amount of material moved with wind speed doubled	5 Increase in amount
6 m/sec		12 m/sec		
7 m/sec		14 m/sec		
8 m/sec		16 m/sec		

Table **8.1** Measuring the effects on sand movement resulting from a doubling of wind speed (see exercise 11)

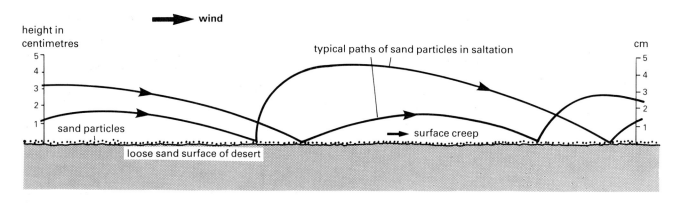

Fig. 8.9 Sand particle movement across the desert surface

11 Study Fig. 8.8 carefully.
a) Why do you think logarithmic scales have been used? (You may wish to refer back to Fig. 5.8 and the accompanying text on p. 106.)
b) You can use Fig. 8.8 to show what effect a *doubling* of wind speed has on sand movement. On a copy of Table 8.1, complete columns 2 and 4 by reading off from the graph. Divide column 4 by column 2 to calculate column 5. What conclusions do you draw?

Since the data in Fig. 8.8 referred to a height of one metre above ground level, the particles being transported would have been carried by the wind in **suspension**. Fig. 8.9 shows what form sand movement might take at ground level. The diagram clearly shows that sand particles tend to move along in a series of leaps, a process known as **saltation**. The larger particles will stay close to ground level and move by **surface creep**. What the diagram does not show is that a sand grain in saltation can shift other particles up to 200 times its own weight! The main effect of such wind action is to remove loose sand from the desert surface, leaving the larger rocks and stones behind. Such landscapes are known as **desert pavements** and are characteristic, as we have seen, of the hammada and reg.

The erosional consequences of this sand movement are very significant. Some rock outcrops show very distinctive **undercutting** and this is usually attributed by scientists to the **sand blast** effect of wind laden with sand. Others believe that chemical weathering may have had an effect at the base of the outcrop, especially as the 'undercutting' is more pronounced on the shadier side where moisture is concentrated.

Fig. 8.10 The Garden of the Gods, Colorado, USA

Desert Processes and Associated Landforms

12 Study Fig. 8.10. In your exercise book draw a sketch of this photograph. Annotate your sketch carefully, indicating how you think the features may have been formed. Try to give some scale to the features (note the mule under the tree). Label the central feature 'rock pedestal'.

As a result of continued movement, the sand grains themselves may be worn down, smoothed off and rounded, a process known as **attrition**.

Occasionally the erosive action of the wind may lead to the development of a shallow hollow which, in time, becomes enlarged into a **deflation hollow** or blow-out. Such hollows are common features of all desert and semi-desert regions; Fig. 8.11 shows how they might develop. Whilst the diagrams show the clear influence of rock structure on the eventual form of the hollow, not all such hollows can be explained in this way. Furthermore, it is unlikely that their formation can always be attributed exclusively to wind erosion.

Deflation hollows vary considerably in size. Most are relatively small, perhaps a kilometre long, a few hundred metres wide and several metres deep. The Big Hollow in Wyoming, USA, is very large at 14.4 km long, 4.8 km wide and 90 m deep. Possibly the largest of all however, is the Qattâra depression in Egypt.

13 Refer to Fig. 8.11 and answer the following.
a) Give details of the desert processes already considered, which are represented by the stippled zone and which lead to the breakdown of the weaker rock.
b) How has structure influenced the form of the hollow? What rock structure is represented?
c) What is the name given to such relief? (Refer back to Fig. 2.13, p. 33.)
d) Apart from undercutting by wind erosion, how else might you explain the recession of the cliffs in the hollow?
e) Locate the Qattâra depression in your atlas. It should be clearly marked. Draw a sketch map of the region and calculate the approximate area of the depression. The lowest point is actually 134 m below sea level.

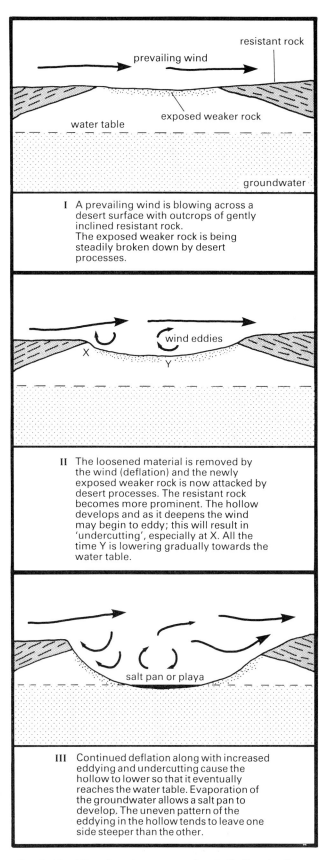

Fig. **8.11** The development of a deflation hollow

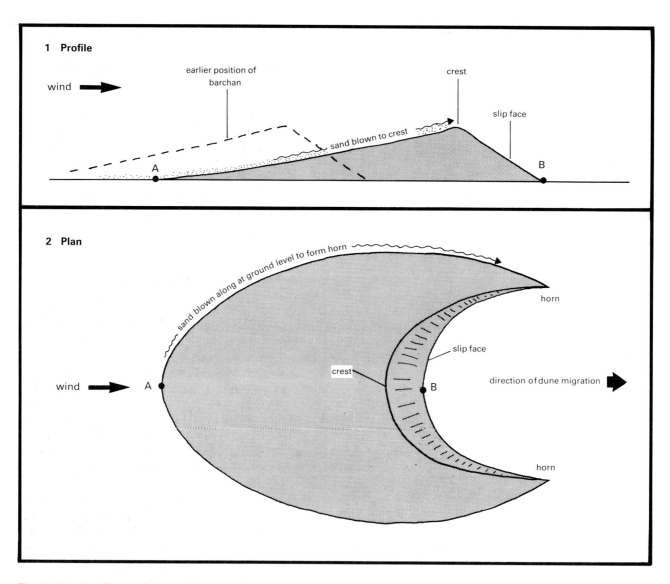

Fig. 8.12 Profile and plan of a barchan dune

Probably the most distinctive of the desert landforms produced by wind action are those resulting from deposition. Sand particles produced by weathering, deflation or abrasion may be transported by the wind to accumulate eventually in one of the extensive **sand seas** or 'ergs' as they are called in the Sahara. These sand seas are constantly shifting under the influence of the wind. Sand accumulations may take a whole variety of different forms, ranging from large dunes to small-scale sand ripples. Any small obstruction may be sufficient to allow loosely blowing sand to build up.

Two main types of sand dune can be identified, their forms influenced primarily by the nature and changeability of the wind direction. **Barchans** are the crescent shaped dunes which develop when the wind blows almost exclusively from one direction. Longitudinal or **seif dunes** tend to develop when wind direction is more changeable. Included amongst other, less significant dune types are the **transverse dunes** which collectively form the typical sand sea; they resemble an ocean swell, with their crests at right angles to the general wind direction.

Fig. 8.12 shows the profile and plan of a typical barchan which commonly occurs in groups. The wind blows sand particles to the dune crest until the angle of slope on the sheltered side of the crest becomes so great that slippage occurs (there is no vegetation binding these particles together). Usually slippage happens when the slope – the **slip face** – exceeds 35°. At ground level wind can sweep sand grains round to form the barchan's very characteristic **horns**. Barchans rarely exceed 30 m in height or 350 m in either width or length. Generally though, they are much smaller and those in Imperial Valley, California, have average heights ranging from 3 to 12 m. Since the supply of sand is continuous, barchans tend to migrate downwind at rates which vary from a few centimetres up to 10 or 15 m per annum.

Desert Processes and Associated Landforms

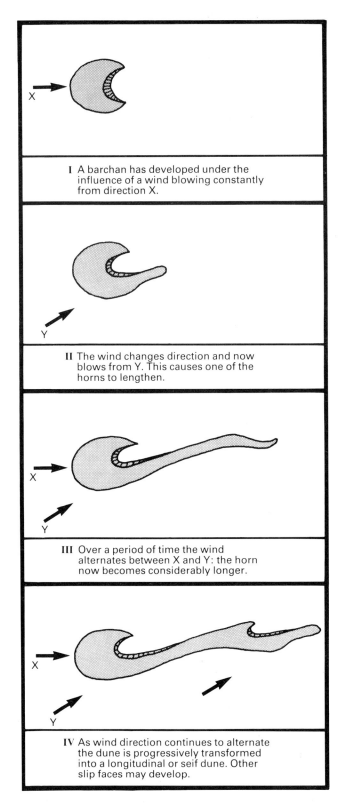

Fig. **8.13** The transformation of a barchan into a seif dune

14 a) Read through the preceding paragraph and using Fig. 8.12 as a guide, draw *scale* diagrams of the profile and plan of a large barchan.
b) Wind eddies (similar to those shown in Fig. 8.11) may be important in undercutting at the foot of the slip face. Mark arrows on your diagram to indicate the wind flow you would expect over a barchan.
c) What effect would such undercutting have on the gradient of the slip face and its stability?

The exact origin of seif dunes is still being debated by scientists, though some believe they may develop from barchans. Fig. 8.13 explains how this transformation might occur under the influence of winds which, over a long period, fluctuate between the directions shown rather than blow predominantly from one direction. Eventually the dunes become completely elongated, so much so that they may start controlling the ground direction of the now inconsistent wind flow. Fig. 8.14 shows that, once established, seif dunes can be maintained by the corkscrew pattern of air flow they create. This renewed wind constancy at ground level may actually allow barchans to develop between adjacent seif dunes.

Fig. 8.15 is a map showing the main trends of longitudinal dunes (seifs) in the relatively flat interior of Australia. Fig. 8.16 is a photograph of one of these dune areas and Fig. 8.17 shows the seasonal prevailing wind conditions in Australia in greater detail than the more generalised maps of Figs. 4.9 and 4.10.

15 a) Study Figs. 8.15 and 8.17 and describe the link between the wind directions shown and the alignment of the dunes.
b) Locate the Simpson Desert in your atlas and, using the information in Figs. 8.15 and 8.17, state the direction in which the dunes shown in Fig. 8.16 are pointing. Justify your answer.
c) Where might the source region of this dune sand be located?
d) Use your atlas to see if there is, in the relief and other features of this assumed source region, any evidence of a desert piedmont. (You may wish to refer back to the map you drew for exercise 9.)
e) Using ideas already mentioned in this chapter and Chapter 6, can you suggest reasons why wind directions might in time become increasingly variable?

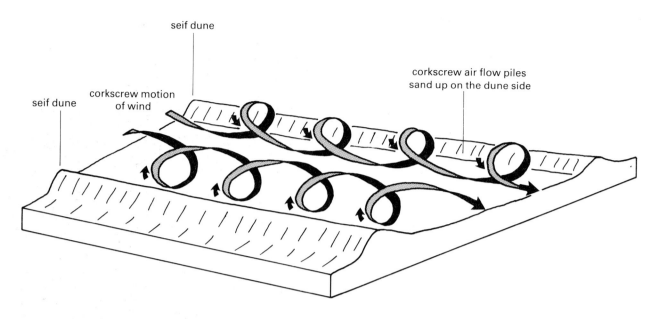

Fig. **8.14** The development of seif dunes by corkscrew air motion

Seif dunes tend in general to be larger than barchans. The very largest may reach nearly 100 m in height, 500–600 m in width and stretch for several hundred kilometres, as can be imagined from Fig. 8.16. Fig. 8.18 is a profile across an actual seif dune in North Africa, located just to the north-east of the Ahaggar massif; you can see the region in Fig. 8.2.

Fig. **8.16** Seif dunes in Simpson Desert, south-east of Alice Springs, Australia

16 Study Fig. 8.18 carefully.
a) What is the orientation of the *long axis* of the seif dune?
b) Use a climatic map of Africa in your atlas to find the prevailing wind in the region. Is it the same as the dune orientation?
c) Explain why one side of the dune should be steeper (see Fig. 8.13).
d) From the profile measure i) dune height, ii) dune width, and iii) ratio of width to height. Is this ratio the same as that for the largest barchans?

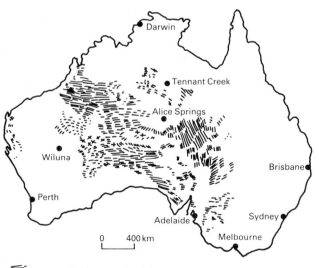

parallel longitudinal dunes

longitudinal dunes of variable trend

Fig. **8.15** Longitudinal dunes in Australia

Desert Processes and Associated Landforms

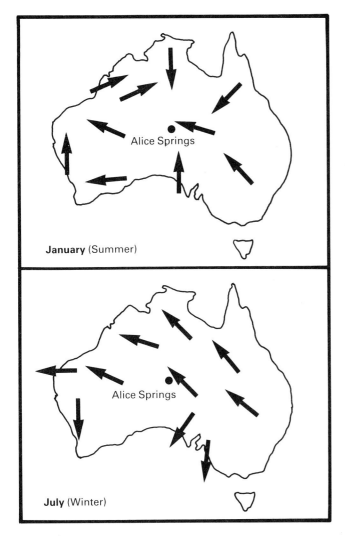

Fig. **8.17** Seasonal prevailing winds in Australia

One characteristic of all desert dunes is the way in which **cross-bedding**, the criss-cross layering of deposits on top of one another, commonly occurs. We have seen how loosely blown sand is often liable to slippage and how other sand may accumulate on top. This process may lead to a typical cross-section through a dune slope looking like Fig. 8.19. Cross-bedding in dunes can help to explain how some of the sandstone rocks in Britain were formed.

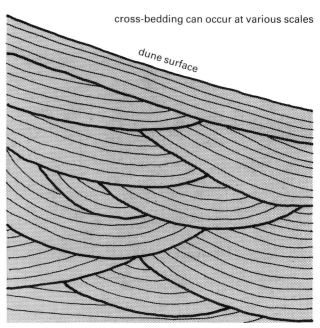

Fig. **8.19** Typical dune cross-bedding

17 Study Fig. 8.20 which shows a cross-bedded sandstone at Dawlish, Devon, formed in Permian times.
a) How long ago was the Permian period? (See Fig. 1.30, p. 22.)
b) What is the present latitude of Devon?
c) If we assume that the world's climate in Permian times was not too different from today's, between what latitudes must Devon have been for this desert-laid sandstone to have formed? (See Fig. 6.3.)
d) What is the approximate difference in latitude of Devon since the Permian?
e) Refer back to Fig. 1.31 to explain why such a difference should exist.
f) Cross-bedding may also occur beneath water. What characteristics would you expect the individual sand grains of this sandstone to reveal under the microscope, to show they were of wind-blown rather than water-borne origin? (See p. 197.)

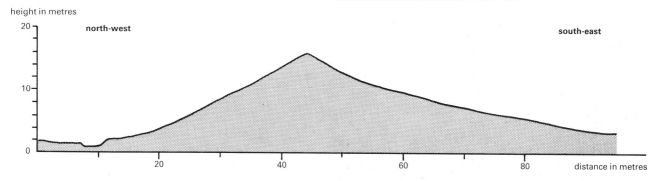

Fig. **8.18** Cross-section through a North African seif dune

Fig. **8.20** Cross-bedding in sandstone, Dawlish, Devon

Finally, we must make a brief mention of **sand ripples**. These are usually no more than a few centimetres high and can develop on all accumulations of sand, including the dunes themselves. Their origins are linked with the process of saltation and it is believed that the distance between ripples (i.e. ripple wavelength) reflects the distance bounced by the average particle in saltation. This in turn is closely related to wind velocity.

Deserts and Man

Deserts, by their very nature, are inhospitable places and support only very small populations. Many of these population groups are traditionally nomadic. The key ingredient for all life in deserts is the availability of water and for humans this really means a permanent and reliable water supply. Even though a location in the heart of a desert may receive virtually no rainfall, such a water supply is still possible. Exotic rivers (see p. 193), rising in much wetter regions, certainly help to support relatively dense desert populations, probably the best example of which is the Lower Nile. Elsewhere small populations may be dependent upon an **oasis**. Fig. 8.21 shows how rain falling in a wetter region on the edge of a desert may seep slowly down permeable strata at a rate often no more than 50 metres per annum, until it reaches the surface, possibly several hundred kilometres away. Two common ways of oasis formation by groundwater or artesian water seepage are shown, each of which may be adequate to support a permanent community.

18 Study Fig. 8.21 carefully.
a) Explain why water is able to reach the surface at the fault and lead to the oasis at X.
b) Suggest reasons for the formation of the depression and the resultant oasis at Y.
c) As well as these natural oases, artificial oases can be developed by drilling boreholes and pumping groundwater to the surface. What problems might result from too many such operations?

Chapter 4 mentioned how some desert soils were naturally fertile and would respond well to irrigation. However, problems may easily arise. One of the most

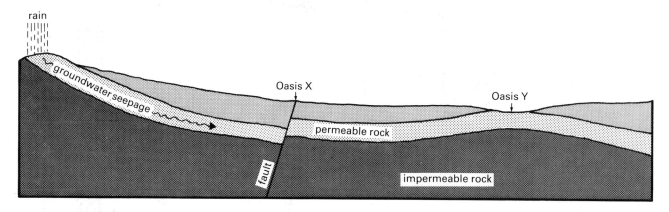

Fig. **8.21** Formation of oases by groundwater seepage

Desert Processes and Associated Landforms

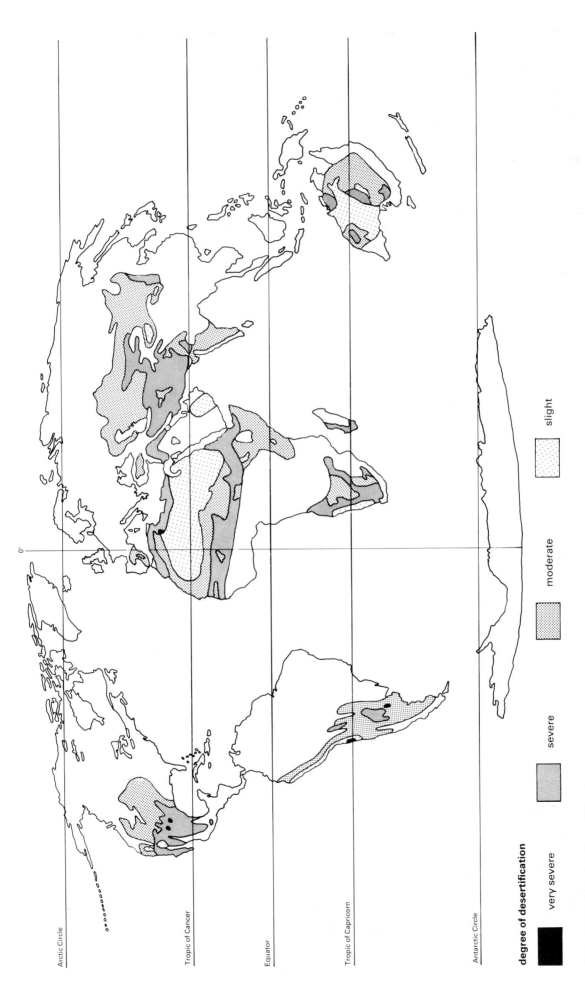

Fig. 8.22 World desertification

serious is **salinisation**, the build-up of salts brought to the surface by the groundwater, which is then evaporated. This concentration of salts leads to infertility and eventually a drop in agricultural production, as is happening in the drier parts of the Prairies today.

Perhaps more serious are the activities of man that actually lead to the creation of deserts. As the human population has been expanding rapidly, so people have required more land for agriculture and have pushed closer and closer to the true desert margins. Where population densities have become too high to be supported by the land (usually semi-desert), overcropping and overgrazing of livestock have occurred. In time this overuse of the land causes a serious drop in its quality and often exposes it to soil erosion (see p. 83), a process known as **desertification**, the man-induced formation of deserts. Though desertification has been going on for at least the last 1000 years it has certainly accelerated this century, being made worse in areas where rainfall is naturally very unreliable. Desertification is not necessarily an irreversible process and careful land management can often reclaim land which is under threat.

Four categories of desertification can be recognised and their locations are shown on Fig. 8.22.

a) **Slight** . . . where there is little deterioration in the vegetation cover, often because it is naturally sparse so man can do little to make it worse!
b) **Moderate** . . . where a significant loss (25%) of agricultural production has occurred following increased soil erosion.
c) **Severe** . . . yields have now dropped by more than 50%. Recovery from this condition will be long and expensive.
d) **Very severe** . . . irreversible, with the development of large dunes or deep gullies. Fortunately such areas remain small.

> **19** Study Fig. 8.22 and Fig. 4.41 which shows the percentage variability of rainfall for the world.
> a) Suggest why the areas of slight desertification appear to experience the greatest rainfall variability.
> b) Do you consider that rainfall variability has contributed to the problems of the areas with moderate and severe desertification? Justify your answer.
> c) Explain what is meant by overcropping and overgrazing (see p. 83).
> d) Describe how soil erosion may occur in semi-desert areas.

In all there is about 50 000 000 km² of land in the world which is experiencing desertification. Table 8.2 shows the percentage breakdown of each of the four desertification categories, whilst Table 8.3 shows the percentage area of each continent actually experiencing desertification. For comparison, the area of the United Kingdom is nearly 250 000 km².

Degree of desertification	Percentage
slight	18.0
moderate	53.6
severe	28.3
very severe	0.1
	100.0

Table **8.2** Percentage breakdown of world desertification for all continents

Continent	Percentage of continent experiencing desertification			
	slight	moderate	severe	very severe
Africa	17.0	22.3	21.9	0.03
Asia	2.7	25.8	10.4	–
Australia	30.2	32.7	8.5	–
North America	0.4	12.6	7.6	0.1
South America	0.7	15.1	1.4	0.2
Europe (Spain only)	–	42.5	10.7	–

Table **8.3** World desertification by continent

Fig. **8.23** Desert region, Western USA

20 a) How many times greater than the UK is the area of the world experiencing some degree of desertification?
b) On an outline map of the world represent graphically the data in Table 8.3. Compare desertification in each continent and explain any differences you recognise.
c) Discuss the main problems that are encountered when trying to reclaim desertified land.

One final point concerns the impact of climatic change on the pattern of desertification. As discussed in Chapter 4, there is increasing evidence that the Earth's climate is getting warmer, possibly towards the super interglacial suggested in Fig. 6.1.

21 Re-read the section on climatic change in Chapter 4 (pages 94-95).
a) Compare Figs. 8.22 and 4.39 (b) and suggest how climatic change might affect desertification in North America.
b) Refer back to Fig. 8.1 and consider what influence a movement of the world's climatic belts might have on desertification.

Workshop

1 Fig. 8.23 shows a desert region in the USA. Describe the landforms featured in the photograph and discuss the processes that may have led to their formation. For any one of these landforms use a representative series of diagrams to construct a *model* of its development.

2 Fig. 8.24 shows part of the Grand Canyon in the USA, cut by the Colorado River; it is 1.6 km deep, 24 km wide and about 300 km long. Fig. 8.25 shows Ayers Rock in Australia which rises nearly 350 m above the surrounding desert plains. For *either* photograph, draw a sketch in your exercise book and annotate it carefully to show the main landscape features and the processes that have led to their formation.

3 A scientist advances the hypothesis that large barchans will advance more slowly than small ones. Do you think that appropriate research would end up proving or disproving such a hypothesis? Justify your answer and consider the problems of such research.

4 Use regional textbooks to describe and discuss the landscape impact of one major irrigation project, e.g. the Central Valley Project in California, the Kara Kum Project in the USSR or the Snowy–Murray Development in Australia.

Fig. **8.24** The Grand Canyon, USA

Fig. **8.25** Ayers Rock, Australia

Additional Map Exercises for the Goring Extract

1 Describe the course of the River Thames and its valley.

2 What is the relationship between the distribution of woodland shown on the map and the relief?

3 With the aid of an annotated sketch map, show how relief and drainage may have influenced the site and layout of South Stoke (square 5983).

4 Compare and contrast the relief and drainage immediately to the north and south of the bridleway between Westfield Farm (578856) and Lollingdon Farm (570854). Attempt an explanation of the differences you observe.

5 Identify the man-made features at 562829, 575817 and 578815. Suggest reasons why they are there.

6 Describe the course of the railway in relation to the relief between 602816 and 584860.

7 In the area to the west of easting 59 and to the south of northing 84, show, with the aid of a sketch map, the relationship between the distribution of farms and the relief.

8 Summarise the main processes that are moulding the physical landscape at the present time over the area covered by the map extract.

Additional Map Exercises for the Aberystwyth Extract

1 Describe the relief, drainage and vegetation to the south of northing 90.

2 From map evidence, suggest reasons why the coastline is used for recreation.

3 With the aid of an annotated sketch map, show how relief may have influenced the site and layout of Talybont (square 6589).

4 Describe the distribution of woodland over the area of the map extract.

5 Compare and contrast the course of the Afon Leri and its valley between 616890 and 635880 and between 635880 and 655892.

6 Write a description of the relief, drainage and vegetation to the north of northing 90. Attempt an explanation of the main features you describe.

7 Compare the course of the B4572 with that of the railway in relation to the relief.

8 Summarise the main processes that are moulding the physical landscape at the present time over the area covered by the map extract.

Acknowledgements

The authors and publisher are grateful to the following individuals and organisations for the use of data and/or ideas:
figs. **1.13**, **1.16**, **1.20**, **1.30**, Holmes D, *Principles of Physical Geology*, Van Nostrand Reinhold (UK) Co. Ltd.; fig. **1.22**, Strahler A N & A H, *Modern Physical Geography*, John Wiley & Sons Inc.; fig. **1.25**, from article by Robert Oliver in the *Journal of Applied Meteorology*, American Meteorological Society; figs. **1.28 (a)** & **(b)**, **2.13**, **2.14**, *British Regional Geology (Scotland Tertiary Volcanic Districts, S W England, Hampshire Basin, Bristol & Gloucester District)*, reproduced by permission of the Director/Institute of Geological Sciences (NERC): NERC Copyright reserved; figs. **1.31**, **3.7**, **3.18**, *Philips Certificate Atlas*, based upon copyright material © 1983 George Philip & Son Ltd.; table **2.1** *SMP Advanced Tables*, CUP; fig. **2.7**, from article by Ward W H in *Geographical Journal*, Royal Geographical Society; fig. **2.26** Sparks B W, *Rocks and Relief*, Longman Group Ltd.; figs. **3.32**, **3.35**, **3.36**, The Controller, Her Majesty's Stationery Office. Crown Copyright reserved; figs. **3.33**, **5.17**, Hilton K, *Process & Pattern in Physical Geography*, University Tutorial Press Ltd.; fig. **3.36**, reproduced from the *Guinness Book of Weather Facts & Feats* (1st Edition); fig. **4.19** (+ **Key**), Eyre S R, *Vegetation & Soils*, Edward Arnold Ltd.; figs. **4.29**, **4.30**, Riley D & Spolton L, *World Weather & Climate*, CUP; fig. **4.32**, *The Daily Telegraph*; fig. **4.33**, Cole & Gorman, *Geography of the USSR*, Butterworth & Co. Ltd.; figs. **4.39 (a)** & **(b)**, **4.40 (a)** & **(b)** *Nature 283*, 17–21 (1980), based on statistics from the Climate Research Unit, University of East Anglia; fig. **4.42**, Haggett P, *Geography: a Modern Synthesis*, Harper & Row; figs. **5.1**, **5.15**, **5.27**, **5.33**, **6.1**, **6.4**, **6.21**, **6.34**, **6.36**, **8.1**, **8.22**, tables **8.2**, **8.3**, reproduced from *The Geographical Magazine*, London; fig. **5.9**, based upon the Ordnance Survey Map sheet SN79, 1:25,000 (1st series) and fig. **6.37** based upon the Ordnance Survey Map, sheets SH70/71, 1:25,000 (1st series), with the permission of

The Controller, Her Majesty's Stationery Office. Crown Copyright reserved; figs. **3.9**, **4.37**, **4.38**, **4.41**, Barry & Chorley, *Atmosphere, Weather & Climate*, Methuen Educational Ltd.; figs. **4.15**, **4.35**, Miller A A, *Climatology*, Methuen & Co.; figs. **1.17**, **1.28 (e)**, Tony Crisp, *Earthquakes, Mountains & Volcanoes* (1979), Thomas Nelson & Sons; fig. **1.19 (a)**, *Teikoku's Complete Atlas Of Japan*, Teikoku-Shoin & Co. Ltd.; fig. **1.19 (b)**, Macdonald G A, *Volcanoes*, Prentice-Hall Inc.; fig. **1.28 (c) & (d)**, Monkhouse F J, *Principles in Physical Geography* (1965), Hodder & Stoughton (formerly University of London Press); fig. **5.21**, Ingle D & Stopp P, *The River Basin*, CUP; fig. **6.10**, Embleton C, *Glaciers & Glacier Erosion*, Macmillan, London, Basingstoke; fig. **6.25**, Professor C A M King; fig. **6.29**, Price R J, *Glacial & Fluvio-glacial Landforms*, Longman Group Ltd.; fig. **6.31**, Embleton C & King C A M, *Glacial & Periglacial Geomorphology*, Edward Arnold Ltd.; fig. **7.3**, based upon fig. 54 of G H Dury's *The Face of the Earth* (Pelican Books, 3rd edition 1971), p. 113, copyright © Dury G H, 1959, 1966, 1971. Reprinted by permission of Penguin Books Ltd. Original material adapted from Sverdrup, Johnson & Fleming; fig.**7.16**, based upon Geography 'A' Level, Paper I, Summer 1979, Oxford & Cambridge Schools Examination Board; fig. **7.18**, Robinson H, *Morphology & Landscape*, University Tutorial Press Ltd. & Geographical Magazine, London; fig. **7.19** based upon an existing map © 1978, George Philip & Son Ltd.; fig. **7.26 (a) & (b)**, Gregory S, *Statistical Methods & The Geographer*, Longman Group Ltd.; fig. **7.27**, Hanwell J D & Newson M D, *Techniques in Physical Geography*, Macmillan, London, Basingstoke; fig. **7.30**, reproduced from the *Times Comprehensive Atlas*, © John Bartholomew & Son Ltd.; fig. **7.31 (b)**, Holmes A, *Principles of Physical Geology*, Van Nostrand Reinhold (UK) Co. Ltd.; fig. **7.31 (a)**, Sparks B W, *Geomorphology*, Longman Group Ltd.; fig. **7.37**, Knowles R & Stowe P W E, *Western Europe in Maps*, Longman Group Ltd.; figs. **8.8**, **8.9**, Strahler A N & A H, *Environmental Geoscience: Interactions Between Natural Systems & Man*, John Wiley & Sons Inc.; fig. **8.11**, Small R J, *The Study of Landforms*, CUP; fig. **8.13**, Robinson H, *Morphology & Landscape*, University Tutorial Press Ltd. (after Bagnold R A); fig. **8.15**, Twidale C R & Foale M R, *Landforms Illustrated*, Thomas Nelson (Australia) Ltd. (after map published in *Australian Geographer* by courtesy of Prof. J N Jennings and the Geographical Society of NSW); fig. **8.17**, Rees H, *Australasia*, Macdonald & Evans; fig. **8.18**, McKee E D & Tibbits G C, *Journal of Sedimentary Petrology*, Society of Economic Palaeontologists & Mineralogists; fig. **8.19**, McKee E D, *Bulletin of American Association of Petroleum Geologists*; fig. **8.21**, Selby M J, *The Surface of The Earth, Volume 1*, Cassell Ltd., Longman Group Ltd. Despite every effort, the publishers have been unsuccessful in seeking permission to reproduce the following: figs. **1.10** (Clarence R Allen), **6.2** & table **5.1**. They ask the authors or their agents to contact them should this book succeed in coming into their hands.

Thanks are due to the following for the use of photographs: fig. **1.11 (a)**, Paul Cleves; fig. **1.18**, George Philip Printers Ltd; fig. **1.24 (a), (b), (c) & (d)**, © Gary Rosenquist/EARTH IMAGES; figs. **1.27 (a)**, **3.23 (a)**, **5.2**, **6.9**, **6.11**, **6.15**, **7.45**, **8.23**, **8.24**, **6.41**, Aerofilms Ltd.; fig. **1.27 (b)**, French Government Tourist Office; fig. **2.1**, Keith Hilton; figs. **2.2**, **5.40**, **6.7**, **6.24**, **7.32**, **7.36**, **8.20**, reproduced by permission of the Director, Institute of Geological Sciences (NERC); fig. **3.8**, Ingrid Holford; fig. **2.3 (b)**, John Topham Picture Library; figs. **3.23 (c)**, **7.13**, Frank W Lane Photo Agency; fig. **3.42**, Satellite Receiving Station, Department of Electrical Engineering & Electronics, University of Dundee; fig. **4.25 (a)**, Forestry Commission; figs. **4.27 (b)** (photographer: G M Wilkins FRPS), **4.25 (b)** (photographer: Walter Rawlings), **7.8 (b)**, Robert Harding Picture Library; fig. **4.25 (c)**, Geoslides; fig. **4.25 (d)**, USDA Forest Service Photo; figs. **4.26 (a)**, **8.6**, Grant Heilman Photography; figs. **4.26 (b)**, **5.16**, **5.26**, Popperfoto; fig. **4.27 (a)**, Novosti Press Agency (APN); figs. **6.13**, **6.28**, Boston Museum of Science photos, by Bradford Washburn; fig. **6.30**, University of Cambridge Committee for Aerial Photography; fig. **6.38**, National Publicity Studios, PO Box 5042, Wellington, New Zealand; fig. **7.14**, Archway Publicity, Tenby; fig. **7.46**, The Controller, Her Majesty's Stationery Office: Crown Copyright reserved; fig. **8.5**, Picturepoint; fig. **8.10**, United States Geological Survey; fig. **8.16** © George Silk (*Life* Magazine)/COLORIFIC Photo Library Ltd.; fig. **8.25**, Australian Information Service Photo by D McNaughton. All photographs, with the exception of those cited above, were taken by the authors.

The authors also wish to express their gratitude to Christine Betts for her assistance in typing the original manuscript, Simon Boyd for his support in launching the project, Diane Short for her careful editorial work, and Tina Dutton for her artwork. Thanks are also due to Elizabeth, Catherine and Timothy Gower for their patience whilst the project has been underway.

Index

Ablation, 136, 147
Abrasion, 140, 141–2, 143, 195
Acidic rocks, 12, 14
Advection, 54, 59
Air masses, 88, 89–91
Alluvium, 115
Altitude of Sun, 48–50
Anemometer, 46
Annual temperature range, 68–71
Anticlines, 8–9, 32–3
Anticyclones, 59–60
Aquifer, 36–7
Arch, 168
Arêtes, 144–5
Aridity, 188, see also drought
Artesian wells, 36
Aspect, 52, 146
Atmosphere: composition of, 52–3
Attrition, 115, 166, 197

Back slope, 32
Backwash, 165
Bahada, 192, 194
Barchans, 198
Barograph, 47
Barometers, 47
Barometric gradient, 59
Basalt, 4, 12, 14, 118, 140
Base-flow, 109
Basic rocks, 12
Batholith, 19
Beaches, 171–7
Beaufort Wind Scale, 46
Bedding plane, 12
Berms, 174
Bifurcation ratio, 105–6
Blow-hole, 168
Boulder clay, 147–51
Bournes, 36
Butte, 193

Caldera, 15
Caledonian mountains, 2, 21
Campos, 86
Canyon, 193
Carbon dioxide, 94–5, 135
Carboniferous Limestone, 34, 37–40
Caves: on coast, 166, 168; underground, 40
Chalk: 13, 26–7; relief, 35–7
Chaparral, 88
Chemical weathering, 25, 37, 190–1
Chernozems, 81
Chinook, 63
Cirques (corries, cwms), 140–2, 145–6
Clay, 12, 26, 80
Cliffs: river, 119; sea, 166–9
Climates: classification, 76–8; fluctuation, 134–5; influence by man, 94–6
Clouds, 54–5, 57
Coasts: classification, 185; deposition, 171–82; of emergence, 184–5; erosion, 166–71; and man, 186; of submergence, 182–4
Condensation, 53, 58–9

Coniferous forest, 80, 92
Consequent streams, 32, 101
Continental drift, 21, 23
Continental regions: cold, 92; tropical, 86–7
Continental shelf, 7
Convection: in atmosphere, 54–6; in Earth, 4, 21
Cool temperate regions, 89–92
Corrasion, 113, 166
Corries, see cirques
Corrosion, 115, 166
Crust of Earth, 3–4
Crustal plates, 5–8
Cuestas, 30–4
Cuspate foreland, 180
Cwms, see cirques
Cyclones, 62–3

Deciduous forest, 80, 91
Deflation: 195; deflation hollow, 197
Deltas, 122–4
Denudation, 24, 98
Deposition: fluvio-glacial, 152–6; glacial, 146–52; marine 171–82; by rivers, 115, 119–24
Depressions, 60–2, 90
Desertification, 87, 204
Deserts: 87; definition, 76, 188; erosion, 191–7; and man, 87, 202–4; weathering, 190–1
Dew, 58
Dew point, 58
Dip of rocks, 8, 32
Dip slope, 32
Discharge: 109; measurement, 110–11
Diurnal range, 50, 87
Doldrums, 71
Dolines, 38
Drainage: basin, 101–9; density, 106–9; diversion, 156–7
Drought, 76, 80, 83, 88
Drumlins, 151–2
Dry adiabatic lapse rate, 54
Dry valleys, 35–6, 158
Dunes: coastal, 180–2; desert, 198–202
Dust storms, 195
Dyke, 19

Earth: history, 20–2; orbit, 48–9, 135; structure, 3–5
Earthquakes, 1, 7
Environmental lapse rate, 50–1
Equatorial regions, 85–6
Equinox, 48
Erg, 188
Erosion: 12, 18, 98; glacial, 139–46; marine, 166–71; by rivers, 98–9, 113–15, 117–18
Erratics, 150–2
Eskers, 154–6
Evaporation, 53, 81
Evapotranspiration, 53
Exfoliation, 25, 190–1
Exotic streams, 193, 202
Extrusive rocks, 12, 13–16

Faulting, 7–8, 9–11, 167
Fault-line, 9, 167

Fault-scarp, 34
Felspar, 12
Ferrel's Law, 59
Firn, 137
Fjords, 183
Flash floods, 191
Flint, 13
Flood plain, 115, 119–22
Fluvio-glacial deposition, 152–5
Fog, 58–9, 60
Folding, 8–9
Fold mountains, 3, 7, 21
Freeze-thaw weathering, 24, 139, 143, 157, 191
Fresh water marsh, 180
Fronts, 58, 60–2, 72
Frost, 58, 60
Frost hollow, 51-2
Frost shattering, *see* freeze-thaw weathering

Gaps, 30–1
Geological column, 21
Geos, 168
Glacial streams, 152
Glaciation: deposition, 146–52; erosion, 139–46; and man, 160–1
Glaciers, formation of, 137–8
Gneiss, 13
Gorges, 40, 118, 130
Graben, 11
Granite, 4, 12, 40
Grassland, 81, 83, 86–7
Greenhouse effect, 94–5, 135
Grykes, 38

Hammada, 188
Hanging valleys, 143
Hardpan, 81
Hawaiian-type volcanoes, 14
Head deposits, 158
Headlands, 164, 168
Headward erosion, 101
Heat budget, 66–7
Heat island, 52
Hercynian mountains, 2, 21
Horse Latitudes, 71
Horsts, 11
Humus, 80
Hurricanes, 62–3
Hydraulic action, 113
Hydrological cycle, 53–4
Hypothesis testing, 42

Ice Ages, 135
Icebergs, 138
Ice: caps, 138; sheets, 138
Igneous rocks: 12; intrusions, 12, 19–20; relief, 40–1
Inselbergs, 192–4
Insolation, 47–50, 66–7
Instability of atmosphere, 54–6, 90
Interlocking spurs, 117
Inversion, 51
Inverted relief, 33
Ironpan, 81
Isobars, 47, 59

Isohyets, 46
Isostasy, 5
Isotherms, 44–5

Joints, 25, 38, 167–8

Kames, kame terraces, 154
Karst, 39
Knick point, 130

Laccoliths, 19
Lagoons, 123, 180
Lakes: cirque (corrie), 144; glacial, 156; oxbow, 122; ribbon, 143–4
Land and sea breezes, 64
Landslides, 29–30
Lapse rates: dry adiabatic, 54; environmental, 51
Laterite, 81, 87
Lava, 14
Leaching, 81
Levees, 121
Limestone: 13; relief, 37–40
Limon, 158
Llanos, 86
Load (of a river), 112–13
Loam, 80
Local winds, 63
Loess, 158
Long profile, 113, 114, 118
Longshore drift, 173, 177–9

Magma, 12
Mantle, 3–4, 5, 6, 7
Maquis, 88
Marine deposition, 171–80
Marine terraces, 184
Marine trench, 7
Marl, 80
Marram grass, 180
Marsh, 180
Mass movement, 29–30
Meanders: 119–22; incised, 130
Mechanical weathering, 24, 190–1
Mediterranean type regions, 88–9
Meltwater streams, 152
Mesa, 193
Metamorphic rocks, 13
Mica, 12
Mid-ocean ridge, 6
Minerals, 12
Misfit stream, 125
Mohorovicic discontinuity, 4
Monsoons, 72, 88
Moraines, 146–8
Mountain climates, 92–3

Nappe, 9
Natural regions, 66, 85–93
Natural vegetation, 78–80
Névé, 137
Normal fault, 10
Nuée ardente, 17
Nutrient cycle, 80

Oases, 87, 202
Ocean currents, 69–71
Oceanic trench, 7
Offshore bars, 179, 182
Onion-skin weathering, 25, 190–1
Orbit of Earth, 48–9, 135
Orogenesis, 7, 21, 22
Outwash plain, 152
Overland flow, 109
Oxbow lakes, 122

Pangaea, 21
Parasitic cone, 14
Paroxysmal eruption, 15
Pediment, 192, 194
Peléean-type volcano, 14, 17
Percolation, 35
Periglaciation, 157–8
Permafrost, 92, 157
Permeability, 18, 27, 35, 108
Piedmont: desert, 192; glacier, 151
Plates, 5–8
Playa, 188, 194
Plucking, 140, 142, 143
Plunge pool, 118
Podzol, 81, 92
Polar Front, 60, 72
Polar regions, 92
Poljes, 38–9
Pot holes: karst, 40; stream beds, 113, 115, 125
Precipitation, 53, 54–8
Pressure: measurement, 47; systems, 59–63; world distribution, 71–3
Prevailing wind, 47
Pumice, 18
Pyramidal peaks, 144
Pyroxenes, 12

Quartz, 12, 25

Radiation, 48, 54
Rainfall: convectional, 54–6, 73, 191; cyclonic, 57–8, 73; frontal, 57–8, 73; intensity, 56, 57, 87, 191; measurement, 45; orographic (relief), 56, 73; regime, 75–6; reliability, 76, 87; world distribution, 73–6
Rain gauge, 45
Rain shadow, 57
Rainwash, 29
Raised beach, 184
Reg, 188, 196
Regolith, 25, 29
Regression lines, 50, 56, 107
Rejuvenation, 118, 125
Relative humidity, 53
Reverse fault, 10
Rias, 183, 186
Ribbon lakes, 143
Riffles, 122
Rift valleys, 11
Rills, 101
Rime, 59

Rivers: capture, 101, 124–5; cliffs, 119, 132; erosion, 98–101, 113, 115; and man, 37, 132–3; terraces, 130; valleys, 115–22
Roche moutonnée, 140

Saltation, 113, 196
Salt marsh, 180
Sand: 12, 80; blast, 196; dunes, 178, 180
Sandstone, 12
Savanna, 86
Scarps, 30–2
Scattergraphs, 50, 56, 107–9
Screes, 24, 139
Sea floor spreading, 6
Sea level changes, 182–4
Seasons, 48–9
Sedimentary rocks, 12–13
Sediments, 4
Shale, 12
Shield volcanoes, 14
Sial, 4
Silica, 12, 14
Sill, 19
Sima, 4
Slate, 13
Sleet, 58
Slopes, 26–30
Smog, 59
Snow, 58, 137
Snow-line, 93, 143
Soil: creep, 29; erosion, 83–4, 86, 87, 88, 204; formation, 80–1; horizons, 80; profiles, 80
Solifluction, 41, 158, 160
Solstice, 49
Solution, 37, 113
Spits, 178
Springs, 35
Stability of atmosphere, 54, 90
Stacks, 168
Stalactites, 40
Stalagmites, 40
Stevenson Screen, 44
Storm beach, 172–7
Strata, 12
Streams: consequent, 32, 101; discharge, 109–11; erosion, 113, 115; exotic, 193; glacial, 152; load, 112–13; order, 104–6; strike, 101; subsequent, 101; transport, 113; velocity, 112–15
Striations, 140
Stump, 168
Subduction zone, 7
Subsequent streams, 101
Subtropical high pressure cells, 71
Swallow holes, 38
Swash, 165
Synclines, 8, 33, 169
Synoptic chart, 60, 61, 90

Tarns, 144
Temperature: annual range, 68; diurnal range, 50, 87; influences on, 47–52, 94–5; inversion, 51; world distribution, 66–9
Tertiary mountains, 3, 7, 9, 21
Thermal Equator, 67

Thermals, 54, 71, 72
Thermometers: dry bulb, 44, 53; maximum, 44; minimum, 44; wet bulb, 53
Through-flow, 109
Thunderstorms, 54
Tides, 171
Tombolo, 179
Tornadoes, 64
Tors, 40
Trade Winds, 73
Tree-line, 93
Tropical rain forest, 86
Tropopause, 51
Truncated spurs, 143
Tundra, 92, 93
Typhoons, 62–3

Uplift, 29, 101

Vales, 30–4
Valleys: dry, 35–6, 158; glaciated, 140, 143; hanging, 143; river, 115–22
Vegetation, 78–80
Vesuvian-type volcanoes, 14
Volcanoes: causes, 6, 7; distribution, 1; erosion, 18; types, 13–18

Wadis, 188, 193
Water cycle, 53
Waterfalls, 118
Watershed, 101, 107
Water table, 35–7
Water vapour, 52–3
Wave-cut platform, 166
Wave refraction, 164
Waves, 162–6
Weather, 43–7, 89
Weathering, 24–5, 190–1
Weather maps, 60, 61, 90
Wells, 36
Whirling hygrometer, 53
Whirlwinds, 64
Winds: anabatic, 93; katabatic, 93; local, 63; measurement, 46; prevailing, 47; world distribution, 71
Wind erosion, 83, 195–7
Wind gap, 125
Wind rose, 46

Xerophytic vegetation, 87

Young fold mountains, 3, 7, 9, 21